Geometry in Ancient and Medieval India

Geometry in Ancient *and* Medieval India

T.A. SARASVATI AMMA

MOTILAL BANARSIDASS
PUBLISHING HOUSE • DELHI
Email: mlbd@mlbd.in • Website: www.mlbd.in

4th Reprint : Delhi, ***2022***
Second Revised Edition : Delhi, 1999
First Edition : Delhi, 1979

ISBN: 978-93-5676-052-3 (Hardbound)
ISBN: 978-93-90064-57-1 (Paperback)

Printed & Published by
Rajendra P Jain *for*
Motilal Banarsidass Publishing House
A-44, Naraina Industrial Area, Phase I,
New Delhi–110 028 (INDIA)
Email: mlbd@mlbd.in
Website: www.mlbd.in

FOREWORD

When the author of the present Thesis came to me to do research, I did not want her to take up any subject in the overworked fields of Alaṃkāra, Vedānta or general literature and wanted to know if she was prepared, to work in fields which were neglected or in which few young scholars were inclined to put forth their efforts. On further enquiry I found that she was qualified in mathematics, having taken her first degree in physics and mathematics and decided that she should specialise in the field of Indian contribution to mathematics, algebra and geometry.

The originality and antiquity of Indian contribution to these branches of science have been questioned by some of the historians of mathematics. For example while it is generally believed that the credit of having discovered the place value and decimal system goes to India, some distinguished modern writers do not accept this. But in the case of geometry, we are on more solid grounds. Not only are the *Śulba Sūtras* earlier in date to Pythagoras but the entire sacrificial system and the fire altars, *vedis* for which the *Śulba Sūtras* were intended, are already presupposed by the Ṛgvedic hymns.[1] The biased view of the ancient Hindu contribution, either for or against, has been aggravated, as observed by an eminent modern Indian scientist,[2] by the inadequate publication of the original documents. Needham says "future research on the history of science and technology in Asia, will, in fact, reveal that the achievements of these peoples contribute far more, in all pre-renaissance periods, to the development of world science than has yet been realised."[3] This study to be useful could be undertaken only by those who have scientific equipment, and if these have the additional

[1]See Macdonell, *History of Sanskrit Literature* and *India's Past* and Bibhuti Bhushan Datta, the *Science of the Śulba*..

[2]See Prof. S. Chandrasekhar, *Astronomy in Science and in Human Culture*, J. Nehru Memorial Lecture, 1968, p. 11.

[3]See Dr. V. Raghavan, Presidential Address, Technical Sciences and Fine Arts Section. XXIst AIOC, New Delhi 1961.

grounding of a knowledge of Sanskrit, the best possible results could be expected. The material available should be interpreted in terms of modern knowledge in the concerned sciences. It is in this respect that work such as the one being introduced here is important.

Dr. Sarasvati has examined ancient Indian geometry as seen in the Vedic period and its *Śulba Sūtras* and in the texts of the classical and post-classical periods of Sanskrit literature, as also in the Jain texts like the *Sūrya*, *Candra* and *Jambūdvīpa Prajñaptis*. The work was recommended for the Doctorate Degree by Judges who were mathematicians and its publication will be an addition to the meagre expositions available on the scientific aspects of Sanskrit literature.

The efforts of the section of the Ministry of Education dealing with the history of Science in India and of the Association for the History of Science and their Journal have been helpful for the development of researches in this field. Special emphasis was laid by the First International Sanskrit Conference held recently by the Ministry of Education, on Sanskrit and Science and Technology and it revealed the talent available for tackling subjects in this area. However it cannot be said that, as in the case of Philosophy, Professors of the different sciences in the Indian Universities have become interested in this subject; as I have pleaded,[1] the history in India of the respective sciences should form a regular complementary part of the study of modern sciences in the Universities and should form legitimate subjects for research degrees for Science graduates.

I hope that the author will continue her investigations in this specialised field and will make further contributions to the elucidation of the Sanskrit literature on mathematics.

Madras
1. 10. 1972

V. RAGHAVAN

[1]See Dr. V. Raghavan, 'The Orient & the West' in *Books*, Journal of the National Book League, London, No. 286, July August 1954, pp. 130-2 and op. cit. Presidential Address to the Section on Technical Sciences and Fine Arts, Delhi Session of the AIOC.

PREFACE

This book is the third in a series of books on Indian Mathematics. The first two, *History of Hindu Mathematics by* B.B. Datta and A.N. Singh, Part I first published in 1935 and Part II published in 1938, concern themselves with Arithmetic and Algebra in Pre-British India. The present book, the author's doctoral thesis, has geometry in the India of the same period as its theme. A similar history of Indian Trigonometry has been compiled by Dr. R.C. Gupta of the Birla Institute of Technology as his doctoral thesis under the guidance of the present author, who has also collected some materials for a history of series Mathematics in India. She hopes to be able to present them in a book form before the research world. A comprehensive history of Indian astronomy is another desideratum to complete the picture of the development of Mathematics in ancient and medieval India.

Indeed the last one should have been the nucleus around which the other sections are to be grouped. For, at least after the Śulbasūtra period, the main developments in Indian mathematics were oriented towards and inspired by the needs of astronomy. The word, *Jyotiṣa* (the science of the luminaries) covered all branches of mathematics. The word '*gaṇita*' (calculations) which combined with *Pāṭi* (calculating board), Bīja (algebraical elements) and *Kṣetra* (field or figure), denotes arithmetic, algebra and geometry respectively has also got an astronomical colouring, since the root '*gaṇ*' has always had a special association with astronomical computations

Geometry, as remarked above, is designated as *Kṣetragaṇita* in most Indian mathematical works. *Kṣetra* means a closed figure whether it be a field or a figure drawn on the calculating board. In the Śulbasūtras and in the Buddhist works *rajju* or *rajjugaṇita* (calculations with the cord) stands for geometrical calculations. It is only very late that we come across the use of the term *Rekhāgaṇita*,[1] calculations connected with the line.

Kṣetragaṇita does not include the calculation of volumes, which is generally given under a separate heading *Khātavyavahāra*.

1. Jagannātha's (c. 1718 A.D.) translations of Euclid's *Elements* is called *Rekhāgaṇita*.

section dealing with excavations. *Rāśigaṇita*, calculations connected with heaps also has some geometrical interest. The present work is mainly based on the *Kṣetra* and *Khāta* sections of available mathematical texts, *Rāśigaṇita* is rarely made use of, since these calculations are usually approximations. Most astronomical calculations being based on geometry, the purely astronomical texts also will yield geometrical material. But such material is not included in this study. In some of the texts like the *Gaṇitasārasaṃgraha*, *Mahāsiddhānta* and the *Gaṇitakaumudī* rules or formulae are given for computing the areas of figures which are not basic geometric figures, but which can be cut up into basic geometric figures like the segment and the triangle. Examples are the figures called *Yava* (barley corn), *muraja* (a sort of drum) and *Śaṅkha* (conch shell). These are also omitted in this study unless they have some special geometrical interest.

The completion of this work as planned is primarily due to the help and encouragement received from my guide, Dr. V. Raghavan, Professor of Sanskrit (since retired), Madras University. I am extremely grateful for his guidance and for gracing this book with his valuable Foreword. I also acknowledge with grateful thanks the help given to me by Prof. T.S. Kuppanna Sastrigal, retired Professor of Sanskrit, Sanskrit College, Madras, Dr. K.Kunchunni Raja, Professor of Sanskrit, Madras University, Prof. C.T. Rajagopal, retired Director, Ramanujam Institute, Madras, Sri Rama Verma (Maru) Tampuran (Joint editor of the *Yuktibhāṣā*), Dr. K.V. Sharma, (now) Reader in Sanskrit, Punjab University, in procuring books and manuscripts and in unravelling the meaning of obscure mathematical passages, and the help of my friends and colleagues Smt. C. P. Parvati, Smt. Helen Barnard and Sri K. R. Prabhakar in correcting the typescript and in typing and proof-reading.

I thank the Government of India for granting me a Humanities Research Scholarship and for sanctioning 50% of the publication cost, though I could not make use of the promised help in time and so forfeited it. I am grateful to the Ranchi University for subsidising this publication in part and to Motilal Banarsidass, Publishers and Book-sellers, for bringing out this work, which, by its very nature, has scant commercial value.

CONTENTS

Chapter I

INTRODUCTION

The mathematical genius of the ancient Indians was mainly computational and led to spectacular achievements in arithmetic and algebra. But the basis and inspiration for the whole of Indian mathematics is geometry. The beginnings of algebra are to be traced to the constructional geometry of the Vedic priests preserved in the *Śulbasūtras*. In later periods also algebra must have leaned on geometry. For, down the ages, Indian mathematicians have shown a predilection for demonstrating algebraical truths geometrically (see Ch. IX). So too the arithmetical operations, multiplication and division were demonstrated geometrically. Trigonometry is, as its name proclaims, the geometry of the triangle. Indian trigonometry which employed sine and cosine chords instead of the ratios has always retained its geometrical character. The study of the Theory of Numbers was really the study of the right triangle and rational rectilinear figures.

The branch of mathematics which received the earliest attention was also geometry. The *Śulbasūtras* (5th to 8th century B.C.) is a manual of geometrical constructions. A geometrical knowledge of this level could not have come into existence overnight. The absence of earlier records has led many scholars to posit large-scale indebtedness to foreign countries, chiefly Babylonia.[1] This is not warranted, though exchange of ideas could have and must have occurred. Even the scant remnants of the Indus Valley Civilisation excavated at Harappa and Mohenjo-daro reveal some acquaintance with geometry. In the words of E. J. H. Mackay, "It is surprising to find that an instrument was actually used for this purpose (i.e. for drawing circles) in the Indus Valley as early as 2500 B.C."[2] Hence it seems more probable that Indian geometry developed gradually

---

[1]M. Cantor, *Vorlesungen Uber Geschichte der Mathematik*, 4th Edition, Vol. I. p. 645.

[2]E.J.H. Mackay, *Further Excavations at Mohenjo-daro*, 1938, p. 222.

into the *Śulbasūtra* stage with the intervening links buried in the sands of time. (India's sands were never so kind to her records as Babylonia's sands have been to her clay tablets.) Even the *Śulbasūtras* may have preserved only a part of the mathematical knowledge of those days, the part that was necessary for constructing the sacrificial altars and for computing the calendar to regulate the performance of sacrifices.

With the sacrificial cult of the later Vedic period waning in influence and the rising ascendency of the cosmography of the Jainas and the astronomy of the Hindus, geometry becomes circle-oriented, so that Indian geometry after the dawn of the Christian era can pertinently be termed chord-geometry. It was studied for the sake of astronomy and along with the rest of mathematics, forms part of astronomical treatises. But this does not mean there was a break in continuity. The *Śulbasūtra* theorem of the square of the diagonal continued as the chief tool in the hands of the astronomer-geometer. The circle-centred nature of Indian geometry was not fully appreciated by some of the Indian mathematicians themselves, the most notable amongst such being Āryabhaṭa II (10th cent.) and Bhāskara II (12th cent.) who were at a loss to understand Brahmagupta's (7th cent.) expressions for the area and diagonals of a cyclic quadrilateral. The amount of geometrical knowledge (as also the knowledge of complicated mathematical series) which went into the derivation of the easy-to-handle rules of computation found in astronomical manuals is also hardly appreciated by the practising astronomer and astrologer. The continuity of the tradition of geometrical knowledge is testified to by the achievements of the Āryabhaṭa school, who not only understand the significance and limitations of Brahmagupta's theorems about quadrilaterals but also give complete proofs of these.

A more serious charge against Indian mathematics in general and geometry in particular is the absence of proofs and demonstrations, so beloved of the Greeks. Yoked as geometry was to practical and astronomical needs, it has remained largely mensurational. In the Indian mathematical texts there is no attempt at any proof or derivation and the earlier commentaries too do not improve the situation much. But luckily there are some commentaries and an independent work meant to be an aid to the understanding of an astronomical text which preserve many elaborate proofs and derivations of complicated mathematical series etc. This shows Indian mathematicians too were not satisfied unless they could prove or derive the results they used. It is also significant that these proofs and derivations are found in commentaries, not in the standard mathematical or astronomical treatises, which satisfy themselves with the enunciation of the results. As in *Āyurveda*, the great teachers were content to provide manuals for easy reference in writing while the explanation and the rationale were left to oral instruction. In fact the break in the oral tradition brought about by foreign invasions may have resulted in significant links in our scientific knowledge being irretrievably lost.

But one has to concede that there was an important difference between the Indian proofs and their Greek counterparts. The Indian's aim was not to build up an edifice of geometry on a few self-evident axioms, but to convince the intelligent student of the validity of the theorem, so that visual demonstration was quite an accepted form of proof.

This leads us to another characteristic of Indian mathematics which makes it differ profoundly from Greek mathematics. Knowledge for its own sake did not appeal to the Indian mind. Every discipline (*Śāstra*) must have a purpose. And since self-realisation and the resulting deliverance from birth and death was the most legitimate purpose of life, those sciences which were supposed to further this purpose directly or indirectly were most assiduously pursued. Thus astronomy was cultivated as an adjunct to the study of the Vedas and the performance of sacrifices with the rest of mathematics as an appendage to it. Hence the paucity of treatises devoted wholly to mathematics. The *Bakhshali* manuscript, the *Pāṭīgaṇita* and *Triśatikā* of Śrīdhara and the *Gaṇitasāra-saṃgraha* of Mahāvīra are the only early works dealing exclusively with mathematics.

Other important developments in Indian mathematics are also the outcome of religious needs. The early geometry of the *Śulbasūtras* grew out of the requirements of altar-construction, mathematical series out of Jaina cosmography and later out of the need for improved methods of astronomical culculations, trigonometry and the geometry of cyclic figures out of the calculations in the celestial circle. Indeterminate analysis, for which the Indians are justly famous, is again intimately connected with astronomical calculations.

Outside this religio-astronomical sphere only the problems of day-to-day life interested the Indian mathematicians. Such were the problems of mensuration, paying for digging and sawing (*khāta* and *krakaca*), barter (*bhāṇḍapratibhāṇḍa*), Rule of Three (*trairāśika*) etc.

The study of permutations and combinations was inspired by literary criticism (the prosody part of it), a necessary accomplishment of a man of refined tastes. Mixing of ingredients in medicines, idol-making and Jaina religious speculations also contributed to the study.

This leaning towards utilitarianism has had unfortunate results. The nonchalance with which the splendid achievements of Greek geometry were ignored, while the pseudo-science of Greek and Babylonian astrology was received with open hands, is perhaps the worst of these. It was only in the 18th century, nearly 2000 years after the active contact of the Indians with the Greeks, that Euclid's *Elements* were translated into Sanskrit[1] and even then perhaps the example of the Arabs provided the inspiration.

General Survey of the history of Geometry and Mathematics in India

In the history of Indian geometry three distinct periods can be discerned :

(1) the pre-Āryan period, the remains of whose civilization have been dug up in Harappa, Mohenjo-daro and other places in the Indus Valley.

(2) The Vedic or *Śulbasūtra* period, and

(3) The post-Christian period.

The first of these, the period of the Indus Valley civilization, is the oldest period in which we can find traces of a civilization in India though eminent scholars like B. G. Tilak and Jacobi would assign a greater antiquity to the civilization of

[1]The translator was Jagannātha, attached to the court of the famous astronomer king, Jayasiṁha.

the Vedic Āryans. The lower limit of this civilization cannot be later than 2500 B.C. The most noteworthy feature of this civilization is well-planned towns for which some knowledge of geometry is indispensable. We also have other evidence of the geometrical sense of this ancient people.[1] A favourite pattern on pottery excavated at Mohenjo-daro is a series of intersecting circles apparently made by drawing a series of vertical lines to divide the surface of the jar into a number of nearly equal panels and then scratching circles with a pair of dividers. Other patterns are the square, the circle, two triangles joined at their apexes, two or four hemispheres with their curved edges towards the middle of the pattern, a series of linked triangles with hemispheres filling up the space and a rectangle with the four sides incurved. "The last of these is a frequently found motif which I have already compared to stretched hide." The motif of the inverted triangles is also fairly common.

We do not yet know what the exact relationship between this city-civilization and the Vedic civilization was. We have to attach some significance to the fact that the commonest motif on the pottery of this period, namely that of the rectangle with incurved sides resembling a stretched hide, is preserved in the shape of the sacrificial altars of the next Vedic period. The most important altar (*vedi*) of Vedic sacrifices, the *Mahāvedi* is of this shape, while its corners are called *śroṇis* (hips) and *aṃsas* (shoulders). This points to a close connection between the geometrical knowledge of the two periods, since the geometry of the Vedic period, as far as we know, existed for the sake of sacrificial altars.

When the poetic vision of the Vedic seers was externalised in symbols, the ritual sacrifice with all its elaborate details came to occupy a pivotal place in their religion. Exact measurements and different geometrical shapes were prescribed for the altars (*vedi*) and fire-places (*agni*) to be used in the sacrifices performed for different purposes. To meet these demands the science of geometry grew up and was codified in the *Śulbasūtras*. But since the sacrificial lore is as old as the Vedas or even older (according to Oldenberg) the geometrical knowledge

[1] *Further Excavations at Mohenjo-daro*, pp. 221-25.

necessary for the construction of the altars and the fire-places must be equally old. The Vedas themselves, as prayers addressed to particular deities, are unlikely to show any traces of geometrical knowledge. Yet we find in them words like *triraśri* and *caturaśri* (*Ṛg Veda* 1, 21, 152, 2), *daśabhuji* (*Ibid.* 1, 10, 52, 11) and *tribhuja* (*Atharva Veda* VIII, 5, 9, 2,). Though these are of doubtful geometrical significance their presence in purely literary works may be an indication of the popularity of geometry in those days.

The *Taittirīya Saṃhitā* of the *Yajur Veda*, the Veda of the performing priest, which gives the measurements of the *Mahāvedi* as 36 as altitude and 30, 24 as parallel sides so that half of 30 and 36 make the sides about the right angle of the Pythagorean triangle 36, 15, 39, must, according to A. Bürk,[1] have been acquainted with the theorem connecting the squares of the sides of a right triangle. The Brāhmaṇas of the *Yajur Veda* contain sporadic accounts of the construction of the *vedis* and *agnis* — the sacrificial altars and fire altars (*Śatapatha Brāhmaṇa* III. 5. 1 & X. 2. 1. 1-3). Finally, along with the codification of the various disciplines for the correct understanding of the Vedas and the performance of the sacrifices, the geometrical knowledge required for the construction of the altars was put together in the *Śulbasūtras*, which form supplements to the *Kalpasūtras* or are incorporated in them. The *Śulbasūtras* cannot be later than the 5th century B.C. and may be as old as the 8th or 9th century B.C. The enunciation in general terms of the theorem of the square of the diagonal and its application to various problems is the most important achievement of this period.

More or less overlapping this period of the sūtras and Brāhmaṇas we find a voluminous Jaina literature which contains a considerable amount of mathematical information. If the correct orientation and measurements of the sacrifical *vedis* were important to the adherents of the Vedic religion, correct calculations in their cosmography were equally important to the Jainas. To them *gaṇita* (calculations or more broadly mathematics) was one of the four *anuyogas*, adjuncts to religious instruction, just as *jyotiṣa* or astronomy was one of the accessories

[1]*Z.D.M.G.*. 1901, pp. 553-55.

to Vedic study. Hence we find mensuration formulae connected with circles, segments, trapezia and trapezoidal solids and rules connected with mathematical series, permutations and combinations scattered through Jaina canonical works, while works like the *Sūryaprajñapti* and *Candraprajñapti* deal exclusively with astronomy as understood by the Jainas. Series mathematics especially seems to have got a great impetus from their cosmographical enquiries. So also the mensurational bent of Indian geometry may be a heritage from them.

The early Jaina works with scattered mathematical material in them are the *Tattvārthādhigamasūtra* more especially its commentary by Umāsvāmin or Umāsvāti, the *Sthānaṅgasūtra*, the *Jambudvīpaprajñapti*, *Anuyogadvārasūtra;* the *Kṣetrasamāsas* etc. As a matter of fact almost all Jaina works, canonical or otherwise, tend to have something of mathematics in them. For the Jainas loved to calculate everything elaborately and precisely, however fantastic their original premises were. But trying to assign dates to these various works is almost a hopeless task. The *Sūryaprajñapti*, which as Thibaut shows,[1] has close affinities with the *Vedāṅga Jyotiṣa*, could not have been much removed from them in time. Moreover Bhadrabāhu, who is said to have written a commentary (*niryukti*) on the *Sūryaprajñapti*, lived 162 years after Mahāvīra i.e. in the 4th century B.C.[2] And the *Sūryaprajñapti* is but an *Upāṅga* (a minor *aṅga*). The *aṅgas* like the *Ācāraṅgasūtras* are likely to be much earlier. Umāsvāmin, commentator of the *Tattvārthādhigamasūtra* and author of the *Jambudvīpasamāsa*, lived in the second century B.C. according to the *Śvetāmbara* tradition and in the 2nd or 3rd century A.D. according to the *Digambara* tradition. Hence we can safely assume that most of these works belong to the centuries before Christ and some may be quite old, since, according to Jaina tradition, Mahāvīra (5th century B.C.) is the last of the Tīrthaṃkaras and Bhadrabāhu the last of the *Śrutakevalins* (those who knew all the 12 *aṅgas*). The *Karaṇagāthās* from

---

[1] *J.A.S.B.* 1880, pp. 107-127 and 181-206.

[2] Jaina School of Mathematics. *Bull. Cal. Math. Soc.* XXI-1929. See also Weber, Sacred Literature of the Jains, *Ind. Ant.* XXI p. 14 onwards,..for the antiquity of the *Sūryaprajñapti*.

which later authors and commentators quote anonymously also appear to be very old.

Few extant works of any mathematical interest belong to the opening centuries of the Christian era. The *Sūryasiddhānta* and the *siddhāntas* summarised in Varāhamihira's *Pañcasiddhāntikā* must have been composed during this period. And these astronomical treatises, unlike the *Vedāṅga Joytiṣa*, presuppose a considerable amount of mathematical knowledge. It is not improbable that treatises exclusively devoted to mathematics were also composed during this period. The *Bakhshali* manuscript (2nd or 3rd century A.D.)[1] may be a representative remnant of such works. Even this manuscript is not of much help in reconstructing the history of Indian geometry, since the geometrical portions it must certainly have contained, have been entirely destroyed.

With the fifth century after Christ we reach surer ground. Āryabhaṭa I (born 475 A.D.) dominated the Indian mathematical world as Plato did the Greek philosophical world, though the extant text of the *Āryabhaṭīya* does reveal some inaccuracies. But the terse language in which it clothes its rules and formulae bespeaks a long tradition of mathematical studies, making it difficult to ascertain what exactly Āryabhaṭa's actual contribution to mathematics is. It is equally difficult to gauge his influence on later mathematicians. Bhāskara I (*c.* 522 A.D.) and Lalla[2] were his followers. Brahmagupta seems to attack him mercilessly in his *Brahmasphuṭasiddhānta* but to be compelled to bow to his superior popularity in his *Khaṇḍakhādyakaraṇa*. The whole host of astronomers which Kerala has produced recognises him as their *ācārya* (master)[3].

---

[1]The *Bakhshali* Mathematics by B.B. Datta, *Bull. Cal. Math. Soc.* XXI, 1929 and The *Bakhshali* Manuscript by A.F. Rudolf Hoernle, *Ind. Ant.* XVII, 1888 p. 33ff. for a discussion of the date. This early date is still a disputed point.

[2]Lalla's date is uncertain. Pt. Sudhākara Dvivedi assigns him to 499 A.D. and S.B. Dixit to *c.*738 A.D. on rather insufficient grounds. K. Balagangadharan in his "A consolidated list of Hindu mathematical works" published in the *Mathematics Student*, XV, 1947, pp. 55-69, assigns *c.* 748 to him.

[3]The author of the *Kriyākramakarī*, a commentary on the *Līlāvatī*, refers to Āryabhaṭa not to Bhāskara as 'Ācārya'.

Many of these like Mādhava of Saṃgamagrāma, Parameśvara of *Dṛggaṇita* fame and Nīlakaṇṭha are astronomers and mathematicians of no mean ability, though to them pointing out mistakes in the master is obviously a sacrilege.

The Āryabhaṭa School of mathematics seems to have had some connection and sympathy with Jaina mathematics. Bhāskara I quotes three Prākṛt verses in his commentary on the Āryabhaṭīya.[1] A second piece of evidence is the large number of the manuscripts of the Jaina Mahāvīrācārya's *Gaṇita-sāra-saṃgraha* found in Kerala. If this hypothesis is correct, the Āryabhaṭa School may be said to have maintained a continuity of mathematical tradition with the Vedic period and the bloom of mathematics in India with Āryabhaṭa ceases to be sudden.

Brahmagupta (628 A.D.) though well-acquainted with Āryabhaṭa's mathematics and astronomy, seems, as already pointed out, to have parted company with the school. His geometry contains some remarkable new theorems about the cyclic quadrilateral. But it is significant that the elucidation and proof of these theorems were undertaken by the Āryabhaṭa School, while Bhāskara II, who closely followed Brahmagupta failed to do so.[2]

Bhāskara I has to be dated between 550 and 628 A.D., round about 574 A.D.,[3] since Pṛthūdakasvāmin makes Brahmagupta later than Bhāskara. In the Āryabhaṭa School Bhāskara's astronomical works (which are really expositions of the *Āryabhaṭīya*)—the *Mahābhāskarīya* and the *Laghubhāskarīya*— are very popular, as is shown by the large number of commentaries on them. But his commentary on the *Āryabhaṭīya* is more interesting to the student of Indian mathematics; for this is the first work in which we come across the geometrical treatment of algebraical formulae which the mathematicians of the Āryabhaṭa School often resorted to. Even series were subjected to this treatment with remarkable success. The only other

[1] *Āryabhaṭīyabhāṣya*. 12806. B. Triv. Uni. Mss. Lib. pp. 5, 10. Also quoted by B.B. Datta in his "A lost Jaina treatise on Arithmetic", *Jaina Siddhānta Bhāskara*, Bhāga 3, Kiraṇa 2.

[2] Vide the chapter on quadrilaterals.

[3] Vide Kuppanna Sastrigal's introduction to his edition of the *Mahābhāṣkarīya* (Govt., Oriental Mss. Lib. publication. Madras 1957).

mathematicians to treat series diagrammatically were Śrīdhara and Nārāyaṇa Paṇḍita, whose method and object differed from those of the Āryabhaṭa School. The diagrammatical treatment of one particular type of algebraic equation, viz. $ax+by+c=xy$ (known as *bhāvitā* in Sanskrit) is met with in Bhāskara II also.[1]

Between Brahmagupta and Bhāskara II we get four treatises devoted to mathematics — the *Pāṭīgaṇita* and *Triśatikā* of Śrīdhara, the *Gaṇitasārasaṃgraha* of Mahāvīra and the *Gaṇitatilaka* of Śrīpati. The Jaina Mahāvīra, whose work is characteristically elaborate, lived in the reign of the Rāṣṭrakūṭa king Amoghavarṣa Nṛpatuṅga who ruled between 814 or 815 and 877 or 878 A.D.[2] Śrīdhara is probably earlier than Mahāvīra though K. S. Shukla places him between 850 and 950 A.D. Śrīpati, who also wrote the *Siddhāntaśekhara* and other astronomical and astrological treatises, flourished in the 11th century. His *Gaṇitatilaka* omits geometry and algebra altogether. To the 10th century belongs Āryabhaṭa II whose astronomical treatise is the *Mahāsiddhānta.* More or less complete lack of understanding of Brahmagupta's theorems about the cyclic quadrilateral, more and more correct approximate formulae for the volume of a sphere, and various approximate formulae for the arc and area of a segment mark these works.[3]

A few Jaina semi-religious works, the *Tiloyapaṇṇatti* (*Trilokaprajñapti*) of Yativṛṣabhācārya and the *Trilokasāra* and *Gommaṭasāra* of Nemicandra (*c.* 980 A.D.) belong to this period. Of these the *Tiloyapaṇṇatti* is perhaps a later redaction of a much earlier work.

Bhāṣkara II (b. 1114 A.D.), wrongly considered the last great mathematician that India has produced, has many original contributions to make, especially in the field of algebra and trigonometry. The accurate formulas for the volume and surface area of a sphere and a close approximation for

---

[1] *Bhāskarīya Bījagaṇitam, Ānandāśrama Granthāvali,* 99, pp. 195-202.

[2] *Gaṇitasārasaṃgraha* ed. by Rangacharya, p. IX.

[3] *The Vateśvarasiddhānta,* whose first volume has been published from Delhi, ought to contain a section on mathematics. But this section has not appeared in print.

the length of an arc in terms of its chord and vice versa are some of his achievements in geometry.

The last of these is a modification of the formula:

$$R \sin \theta = \frac{R\theta\,(180-\theta)}{\frac{1}{4}\{40500-\theta\,(180-\theta)\}}$$ given by Bhāskara I in his *Mahābhāskarīya*.

Nārāyaṇa Paṇḍita, son of Nṛsimha and author of the *Gaṇitakaumudī* (1356 A.D.) is sometimes confused with Nārāyaṇa, commentator of the *Līlāvatī*. The *Gaṇitakaumudī* gives the impression of being a full and detailed treatment of the mathematical knowledge of Nārāyaṇa's age. A few of his theorems —three and only three diagonals are possible for a cyclic quadrilateral with given sides, the area of a cyclic quadrilateral is the product of the 3 diagonals divided by twice the diameter— are not met with in earlier works.[1] In his treatment of cyclic quadrilaterals, Nārāyaṇa has some affinity with the Āryabhaṭa School. But his *śreḍhīkṣetras* (diagrams of series) do not bear the stamp of that school, nor does he seem to be familiar with the very close approximations to the value of π which the school had arrived at probably by about the same time.

Samgamagrāma Mādhava, Parameśvara, Nīlakaṇṭha, Putumana Somayājin and the author of the *Kriyākramakarī*, an elaborate commentary on Bhāskara's *Līlāvati* and the author of the *Yuktibhāṣā* which purports to be an exposition in Malayālam of Nīlakaṇṭha's *Tantrasaṃgraha* — all these belong to a period of intensive astronomical studies in Kerala from the 14th to 17th centuries A.D.

A method of integration based on subtle geometrical analysis and ingenious methods of summing up series, used to get infinite series for π, $\sin\theta$, $\cos\theta$ and the arc, is the most important achievement of this period. This method was also successfully employed for finding the volume and surface area of a sphere. Strictly geometrical proofs for Brahmagupta's theorems, trigonometrical identities, etc. are also found in the works of this period.

---

[1]Caturveda Pṛthūdaksvāmin, the commentator of Brahmagupta (who lived in the 10th century A.D.), is said to have proved Ptolemy's theorem.

From the references scattered through the works of others, one gathers that the remarkable developments in the geometry of the circle found in this period are the achievements of Mādhava, though his extant works, the *Veṇvāroha* and the *Candravākyāni*, do not mention these developments. Mādhava lived in the latter half of the 14th century.[1] Parameśvara who is said to have spent 55 years observing the sky and then corrected the astronomical elements to tally with his observations in 1430 A.D., has a number of astronomical works to his credit. But unlike the *Kriyākramakarī*, his commentaries on the *Āryabhaṭīya* and the *Līlāvatī* do not introduce much new mathematical material. Nīlakaṇṭha's (1443-1545 A.D.) commentary on the *Āryabhaṭīya* is much more elaborate and mathematically interesting. His *Tantrasaṃgraha* also, though mainly astronomical, contains all the new developments in mathematics.[2]

Putumana Somayājin's *Karaṇapaddhati* of uncertain date (but probably of the early 18th century) seems to be of the nature of a compendium. On the other hand the *Yuktibhāṣā*, as the only work which gives the rationale and poof or derivation of all the theorems and formulae then in use among the astronomers of Kerala, is a unique work. The author of the work is, according to K.V. Sharma,[3] Jyeṣṭhadeva, who lived between A.D. 1475 and 1575. And from the manner of his writing he is no innovator in the line of demonstration, derivation and proof. In fact, his remarkable expositions recur in the *Kriyākramakarī*[4] (in part) and in a Sanskrit commentary on the *Tantrasaṃgraha* preserved in the Sanskrit College Library

[1]K.V. Sharma's introduction in the *Veṇvāroha* ed. by K. Achuta Poduval, Sri Ravi Varma Sanskrit Series. No. 7, pp. 7-8.

[2]The text of the *Tantrasaṃgraha* published from the Triv. Mss. Lib. does not contain these mathematical portions, which are found in another manuscript of the *Tantrasaṃgraha* with a Malayālam commentary, a transcript of which Sri Rama Varma Maru Thampuran kindly lent to me and a *Tantrasaṃgrahavyākhyā* in Sanskrit in the Trippūṇithura Skt. College library No. 275 (572).

[3]Jyeṣṭhadeva and his identification as the author of the *Yuktibhāṣā* by K.V. Sharma, *Adyar Library Bulletin* Vol. XXII, pp. 35-40.

[4]In the *Kerala Sāhitya Caritra* the *Kriyākramakarī* is given as an alternative name for Nārāyaṇa's *Karmapradīpikā*, another commentary on the *Līlāvatī*. But the two commentaries are not the same. The *Kriyākramakarī* is

at Trippunithura.[1] The author of the former is probably Śaṅkara Vāriyar (1556 A.D.), a pupil and commentator of Nīlakaṇṭha. Hence it is not unreasonable to believe that these proofs were known in the Āryabhaṭa School at least from the time of Mādhava, if not earlier.

The *Sadratnamālā*, a later work composed by Śaṅkaravarman, contains all these results without the proofs. No new ground was explored after this, chiefly because the needs of astrological astronomy were already satisfied. Secondly, the contact with the West, which, with the brute force of fire-arms overpowered the superior skill and artistic perfection of the military techniques of the native warriors, had a powerful degrading effect on all indigenous arts and studies.

---

full and elaborate, the *Karmapradīpikā* seems to be indebted to Parameśvara's commentary on the *Līlāvatī*.

[1]Transcript No. 275 (572). This is of unknown authorship. But at the end of every chapter there occurs the verse:

इत्येष परक्रोडावासद्विजवरसमीरितो योऽर्थः ।

स तु तन्त्रसंग्रहस्य···अध्याये मया कथितः ।।

(with the chapter number before the word *adhyāya*).

The commentator has derived his knowledge from a Brahmin of Parakroḍa, who may be Jyeṣṭhadeva of Paraññot (Sanskrit—Parakroḍa) family, who is the author of the *Yuktibhāṣā*.

CHAPTER II

ŚULBASŪTRA GEOMETRY

2.1. As much of the geometry of the Vedic period as is required for the construction of the altars (*vedi*) and fire-places (*agni*) prescribed for the obligatory (*nitya*) and votive (*kāmya*, performed to attain specific ends) rites, is contained in the *Śulbasūtras*. These form part of the vast literature, designated as *Kalpasūtras,* attached to the Vedas as one of the six *V dāṅgas* (limbs of the Veda). Louis Renou lists 8 *Śulba-sūtras,*[1] *Laugākṣi, Mānava, Varāha, Baudhāyana, Vādhūla, Āpas-tamba, Hiraṇyakeśin* and *Kātyāyana.* With the *Maitrāyaṇa* mentioned by Datta[2] the number reaches nine. Those of Baudhāyana, Āpastamba, Kātyāyana and Manu form separate treatises; the rest are chapters or parts of chapters of the corresponding *Śrautasūtras.* All these belong to the *Yajur Veda,* the *Kātyāyana* to the *Śukla* and the rest to the *Kṛṣṇa Yajur Veda.* Since all these *Śulbas* are attached to *Śrautasūtras* B.B. Datta surmises that there must have been many *Śulbasūtras* attached to each of the 1131 or 1137 Vedic *Śākhās* with their own *Śrautasutras.*[3] But it seems more likely that the *Śulba-sūtra* sections were confined to the *Śrautasūtras* of the *Yajur Veda*, the Veda specially designed for the performance of the sacrifice.

2.2. *The Date of the Śulbasūtras*

Obviously the geometry of the *Śulbasūtra* has developed with the demands of the construction of the sacrificial *vedis,* and therefore it must be very old. According to Oldenberg, the three most primitive, *agnis,* the *Gārhapatya, Āhavanīya* and *Dakṣiṇāgni* are older than the *Ṛg Veda.*[4] There is a likelihood of the *Mahāvedi* being known to the Indians of the Indus Valley

---

[1] *Vedic India* by Louis Renou, translated from the French by Philip Spratt. Susil Gupta (India) Private Limited, Calcutta 12. 1957. pp. 43 and 51.

[2] B.B. Datta, *The Science of the Śulba*, p. 2.

[3] *Ibid*, p. 1 and foot-note 2 on the same page.

[4] *S.B.E.* XXX, p. IX. Also the Ṛg Veda refers to fire as *trisadhastha*

Civilization even.[1] M. Cantor and A. Bürk recognise as incontrovertible the fact that the Pythagorean theorem was known in India at the latest in the 8th century B.C.[2] The *Kāmyāgnis prauga* (triangle), *ubhayataḥprauga* (rhombus), *ratha-cakra* (wheel), *droṇa* (trough) and *śmaśāna* (cemetery) find mention in the *Taittirīyasamhitā*.[3] The *prācī* (the exact east-west line) and the exact measurements of the *Saumikīvedi* are also mentioned in the same *samhitā*. The *Śatapatha Brāhmaṇa* gives detailed instructions for the construction of the most complicated *agni*, the *Vakra-pakṣa-vyasta-puccha-śyena* (the falcon with curved wings and spread tail). All this shows that the construction of the sacrificial altars with all due precision was a common practice during the Brāhmaṇa and even the Samhitā periods.

We do not know when exactly the codification of the rules for the construction of the *vedis* began, nor do the existing *Śulbasūtras* lend themselves to exact dating. A.B. Keith places the *Āpastamba Śrautasūtra* in the 4th century B.C. and the *Baudhāyana Śrautasūtra* in the 5th century B.C. Bühler would have the *Āpastamba Śrautasūtra* 150-200 years earlier than Pāṇini. A. Bürk accepts Bühler's view. The language of the *sūtras* alone can give us some clue to their age, and the most we can say from their language is that they are pre-Pāṇinian. Keeping in view the dates assigned to Pāṇini by later researchers, one feels *c*. 800 B.C. is a more probable date for the codification of the *Śulbasūtras*.

2.3. *The term Śulba*

The word *śulba* does not occur anywhere in the body of the *Śulbasūtras* except in the metrical supplement of the *Kātyāyana Śulbasūtra*. The word is supposed to have been derived from

(remaining in three places), (RV. 1.67-5), Also सद्मेव धीरः सम्माय चक्रुः refers to expert measurements of the fire place (Vide *Z.D.M.G.* 1901).

[1]Vide Ch. I, p. 4 above.

[2]*Vorlesungen uber Geschichte der Mathematik*, Vol. I, 4th Edn. p. 636 and *Z.D.M.G.* 1901, pp. 553-55.

[3]A Bürk in the introduction to his edition of the *Āp. Śl. Śū.* brings up all these references (*Z.D.M.G.* 1901, p. 548).

the root *śulb* or *sulv* to measure, when *śulba* will mean a measuring tape or cord. But the *sūtras* themselves use the word *rajju* not *śulba* in this sense. The *Kātyāyana Śulbasūtra* opens with रज्जुसमासं वक्ष्यामः. (We will expound the manipulation with cords). Later writers too never use the word *śulba* except in the sense of the *Śulbasūtras*.

2.4. *Analysis of the contents*

The geometry of the *Śulbasūtras* is primarily constructive, though they occasionally notice and formulate some of the geometrical truths involved. The altars and fire-places had different shapes, all geometrical (even the *vakrapakṣavyastapuccha* is not ungeometrical). The orientation, shapes and areas of these had to be strictly correct, this correctness being as important as the correct pronunciation of the Vedic *mantras*. Hence accurate geometrical methods of construction were evolved. Most often the geometrical truths underlying these constructions were left unenunciated. Hence the geometrical contents of the *Śulbasūtras* can be broadly divided into three categories—(1) theorems expressly stated, (2) constructions, and (3) the geometrical truths implied in these constructions but not stated as such.

2.5. *The Theorem of the square of the diagonal*

To the first category belongs the theorem popularly associated with the name of Pythagoras (*c*. 540 B.C,) connecting the square on the hypotenuse or diagonal of a rectangular triangle with the sum of the squares on the sides containing the right angle. Perhaps the first statement of this theorem in its most general geometrical form is ancient India's most important contribution to the development of mathematics. It is true that most ancient peoples knew and used the right triangle 3, 4, 5[1] for getting a right angle and the Babylonian records contain a list of Pythagorean numbers.[2] But the full

[1]The Chinese Nine Sections (*c*. 1100 B.C.) mentions this triangle and the Kahun Papyrus of Egypt (*c*. 2000 B.C.) refers to four sets of numbers forming right triangles (D.E. Smith. *History of Mathematics*, Vol. II, p. 293).

[2]O. Neugebauer, *The Exact Sciences of Antiquity*, p. 35.

geometrical significance of the theorem, that the sides of any right-angled traingle will exhibit this relationship amongst them, was perhaps first realised by the altar-building Vedic priests. The German mathematicians A. Bürk and M. Cantor discuss the question in detail and come to the conclusion that the theorem was known in India at the latest by the 8th century B.C. i.e. the date of the oldest Śulbasūtra that, of *Baudhāyana.* The *samhitāś* of the Black Yajur Veda and the *Śatapatha Brāhmaṇa* give 36 units as the length of the *prācī* (east-west line) or *pṛṣṭhyā* (the line of symmetry) of the *Mahāvedi* and 30 units as one of the north-south sides, the *prācī* and half the side thus making the sides containing the right angle in a rational right angled triangle, viz. 36, 15, 39. In the *Śatapatha Brāhmaṇa* (X. 2.3.7) mention is found of increasing the size of the *Vedi* fourteen-fold to accommodate the *Ekaśatavidhāgni* (the fire-place for the 101st performance of the sacrifice) for which operation a knowledge of the theorem of the square of the diagonal is indispensable.[1] Hence it is highly probable that the theorem of the square of the hypotenuse was known in India much earlier than the *Śulbasūtra* period. Tentative evidence for a very much earlier date even can be brought forward, if the term *triṣadhastha* (residing in three abodes) as applied to the fire, has reference to the three *agnis, Gārhapatya, Āhavaniya* and *Dakṣiṇa,* because the construction of the *Dakṣiṇāgni* requires a knowledge of this theorem. Though the *Śulbasūtras* themselves speak only about the orientation of the 3 fires, the commentators supply the mode of construction, which involves drawing a square with the *dvikaraṇī* of a *piśīla*[2] as the side.[3]

[1]A.B. Keith (*J.R.A.S.* 1910 pp. 519-21) maintains that the mention of the numbers 36 and 15 as representing the sides about a right angle is no evidence of the knowledge of the theorem of the square of the hypotenuse on the part of the Indians of the *Samhitā* period. The evidence of the *Śatapatha Brāhmana* passages enjoining the increment of the *vedis* or altars for each subsequent construction by one *puruṣa,* is also dismissed with the remark that the increment is most probably in the side not the area. But this is disregarding the tradition preserved in all the *Śulbasūtras.*

[2]The *piśīla* is a unit of length defined variously as the arm with folded palms or the length of the two arms extended.

[3]See Kapardin's and Karavinda's comments on *Āp. Śl.* II. 4.

The *Śulbasūtras* exhibit a thorough familiarity with the properties of the right triangle, or rather the properties of the sides and diagonals of figures with right-angular corners. Many rational right triangles are mentioned like :

15, 36, 39 (*Āp. Sl.* V. 2; *B.Sl.* I.49)
3, 4, 5 (*Āp. Sl.* V. 3; *B.Sl.* I.49)

and the latter multiplied by 4 and 5

5, 12, 13 }
12, 35, 37 } (*Āp. Sl.* V. 4; *B. Sl.* I. 49)
7, 24, 25 (*B. Sl.* I. 49)
72, 96, 120 }
40, 96, 104 }
$2\frac{1}{2}$, 6, $6\frac{1}{2}$ } (*M. Sl.*[1])
$7\frac{1}{2}$, 10, $12\frac{1}{2}$ }

Besides these the irrational right triangle 1, 1, $\sqrt{2}$ and the approximate right triangles[2]

36, 90, 97
40, 40, 56
5, 6, $7\frac{5}{6}$
and 4, 4, $5\frac{2}{3}$

are used in the *Śulbasūtras* for constructing right angles. Secondly the fact that the square on the diagonal (*akṣṇayārajju*) of rectangles and squares combines the squares on the two sides is used to find geometrically the side of a square of any given area, whether the number representing the area is a perfect square or not; i.e. to evaluate square roots and surds geometrically and for combining and subtracting squares to yield squares again.

The actual enunciation of the theorem in the *Śulbasūtras* is not with respect to the right triangle but with respect to the sides and diagonals of squares and rectangles.

दीर्घस्याक्ष्णयारज्जुः पार्श्वमानी तिर्यङ्‌मानी, च यत् पृथग्भूते कुरुतस्तदुभयं करोति

(*Āp. Śl.* I. 4)

(The diagonal cord of a rectangle makes both (the squares) that the vertical side and the horizontal side make separately) and

[1] As quoted by Datta in the "*Science of the Śulba*, p. 123.
[2] Vide The *Mānava Śulbasūtram* by N.K. Mazumdar. Jour. Dep. Let. Cal. VIII. 1922.

चतुरस्रस्याक्ष्णयारज्जुर्द्विस्तावतीं भूमिं करोति । (*Āp. Śl.* I. 4)

(The diagonal cord of a square makes double the area). The two parts in the enunciation perhaps indicate two steps in the discovery of the theorem.

Many have speculated on the possible mode of discovery of this theorem. Amongst them the most noteworthy are Thibaut, Bürk and Datta. Thibaut thinks that the Indians must have been "observant of the fact that the square on the diagonal is divided by its diagonals into four triangles each of which is equal to half the first square. This is at the same time an immediately convincing proof of the Pythagorean proposition as far as squares or equilateral rectangular triangles are concerned" (*J.A.S.B.*, 1875, p., 234).[1] That is, the diagram which suggested the theorem was like fig. 1. Bürk substantially agrees with this but would find the inspiring figure in the square *ātman* or body (fig. 2) of the *caturaṣra-śyenacit-agni* (the fire-place in the form of the square falcon). The *ātman* is 4 square *puruṣas* in area and it is to be made by conjoining 4 squares each of area one square *puruṣa*. If one joins the diagonals of these small squares as in figure 2, a square consisting of 4 triangles results, and each of these triangles = ½ the one *puruṣa* square. Hence the theorem.

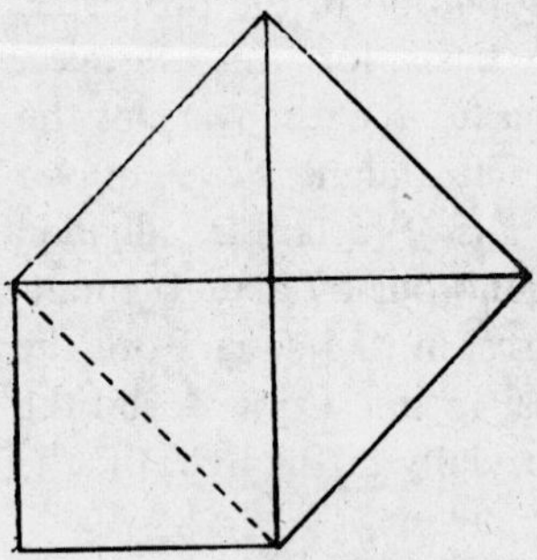

Fig. 1

B. Datta's hypothesis differs but slightly from Bürk's. According to him the construction of the *Paitṛkī Vedi* as directed by Kātyāyana would immediately suggest the theorem. For the prescription is:

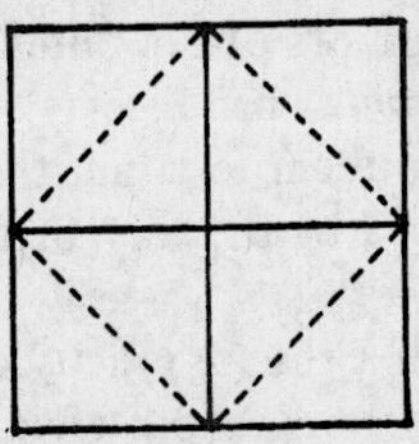

Fig. 2

पैतृक्यां द्विपुरुषं चतुरश्रं कृत्वा करणीमध्येषु शंकवः, स समाधिः । (*K. Sl.* II. 6)

(In the construction of the *Paitṛkī Vedi* make a square of area two square *Puruṣas* and have nails (for the corners of the *Vedi*)

[1]Also quoted by **A. Bürk.**

at the middle points of the sides. That is the construction). The diagram for the construction, which will be the same as fig. 2, yields the theorem as an inference without any alteration or addition. As such, this hypothesis seems to be the most plausible.

For the theorem as applicable to the general rectangle or to the right triangle, most authorities including M. Cantor have surmised a numerical origin. The early mathematicians making figures with pebbles must have noticed that 9 and 16 are square numbers and when the two are combined to give 25 pebbles, the new pile of 25 pebbles also is capable of being arranged as a square. The superior antiquity of the knowledge of Pythagorean numbers as compared to a recognition of the properties of the right triangle in Babylonia is the basis for this surmise. Thibaut accepts the same explanation for the discovery of the theorem in India, but adds that the Indians might have drawn the squares on the sides and diagonals of rectangles, divided these squares into unit squares, and found out by actual counting that the number of unit squares in the square on the diagonal is the sum of the unit squares in the squares on the sides. Such a discovery is plausible especially in the light of the direction, तान् समस्येत् in Kātyāyana's.

यावत्प्रमाणा रज्जुर्भवति तावन्तस्तावन्तो वर्गाः भवन्ति, तान् समस्येत् ।

(III. 3)[1]

(There will be as many times as many small squares[2] as there are units in the cord. These should be added together). That is, area was determined by dividing the figure into unit squares and counting these squares. And to the authors of the Śulbasūtras, ever busy with figures, lines were *karaṇis*, producers of areas and mere curiosity might have driven them to count the number of unit squares in the square produced by the diagonal of a rectangle.

Bürk and Datta turn to the enlargement of a square practised by the Vedic priest as the source of the theorem. A particular

[1] *Āpastamba* (III. 7) has the same *sūtra* with slight changes in the wording.

[2] I have taken *varga* to mean square though Thibaut, Bürk and Datta take the word to mean group. The repetition of the word *tāvantah* and the use of the word *varga* in the sense of square, common in later mathematics, make this interpretation more plausible.

application of such enlargement occurs in Baudhāyana's rule for the construction of the *Sāra-ratha-cakra cit*, i.e. the fire-place in the shape of a wheel with spokes. Bricks are to be made in such proportion that 225 of them together make up the ordained area of the fire-place. To these, 64 similar bricks are to be added and the total (that is 289) is arranged as a square. For this, a square of 256 bricks is to be made first and the remaining 33 are to be put round it. Whatever the necessity for this strange direction, when 289 bricks could be arranged straight away as a square with 17 bricks in each side, this makes it clear that enlarging squares by adding gnomons was a well-known practice. And if the Indians had begun with a square of side 1, adding gnomons to it to get squares with side 2, 3,...... ., one by one they would easily have noticed $3^2 + 4^2 = 5^2$, $15^2 + 8^2 = 17^2$ etc. To explain the geometrical character of the enunciation of the theorem in the *Śulbasūtras*, Bürk [1] fancies that the Indians might have put the squares 3^2, 4^2, 5^2 as in Fig. 3 and thereby discovered the rational right triangle 3, 4, 5.

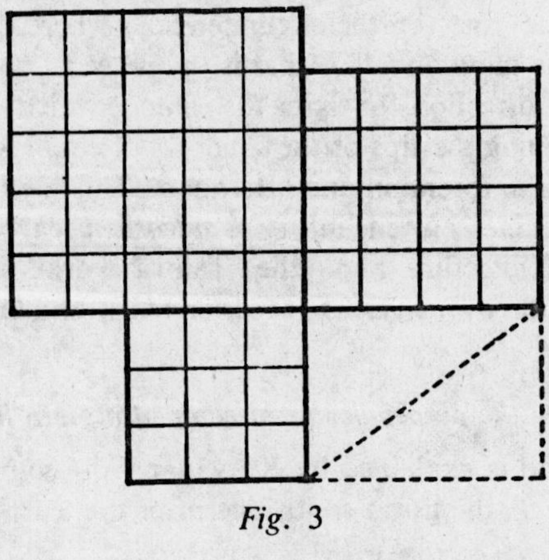

Fig. 3

To speculate on whether the Indians had a proof for the theorem or what the proof could have been is idle. The Śulbasūtras, our only means of knowing what the condition of mathematics then was in India, are only practical manuals for

[1] *Z.D.M.G.* 1901, pp. 567-68 and Datta-*Science of the Śulba*-pp. 125-127.

the construction of the altars. Proofs are outside their scope. Very likely they had proofs orally transmitted to the enquiring student.

The constructions dealt with in the Śulbasūtras comprise the construction of the east-west line; of perpendiculars; of squares, rectangles, and trapezia; and of triangles and rhombi equal in area to a given square; conversion of squares into rectangles and vice versa; of squares into circles and vice versa.

2.6. *Determining the east-west line*

This was preliminary to the construction of all the altars and fireplaces described in the Vedic literature. But it is only Kātyāyana and Manu that give the details of the procedure. *Baudhāyana and Āpastamba* take the *prācī* or east-west line for granted.

समे शंकुं निखाय शंकुसम्मितया रज्वा मण्डलं परिलिख्य यत्र लेखयोः शंक्वग्रच्छाया निपतति तत्र शंकू निहन्ति सा प्राची ।

(*K.Sl.* I. 2)[1]

(Fixing a pin (or gnomon) on level ground and drawing a circle with a cord measured by the gnomon, he fixes pins at points on the line (of the circumference) where the shadow of the tip of the gnomon falls. That is the *prācī*.)

This is the direction for fixing the *prācī*, which is therefore the line joining the tips of the shadows of equal length cast by an object in the forenoon and afternoon. This method of fixing the east-west line is given in the *Tantrasamuccaya* and other works on architecture and the Tantra works dealing with the construction of *maṇḍapas* (sheds or halls) and *kuṇḍas* (sacred fire-pits).

2.7.1 *To draw the perpendicular bisector of a given line*

a. The method is explained by Kātyāyana in connection with fixing the *udīci*, the north-south line after the east-west line is fixed.

तदन्तरं रज्वाभ्यस्य, पाशौ कृत्वा, शङ्क्वोः पाशौ प्रतिमुच्य, दक्षिणायम्य मध्ये शंकुं निहन्ति । एवमुत्तरतः सोदीची ।

(I. 3)

[1]The *Mānava Śulbasūtra* (p. 2) gives the same method. Cantor mistakenly says (*Vorlesungen über Geschichte der Mathematik*, 4th Ed. Vol. I, p. 637) that the *Śulbasūtras* are silent over the method of determining the *prācī*.

(Doubling the distance between them (the end pins) on a cord and making ties one fixes the ties on the pins, stretches (the cord) to the south and strikes a pin at the middle point (of the cord). Similarly to the north. That is the north-south line).

This is the same as the modern method of drawing the perpendicular bisector of a line. Only, instead of drawing intersecting arcs to get two points equidistant from the ends of the line, isosceles triangles are drawn on either side of the line with the line as the base and their vertices are joined.

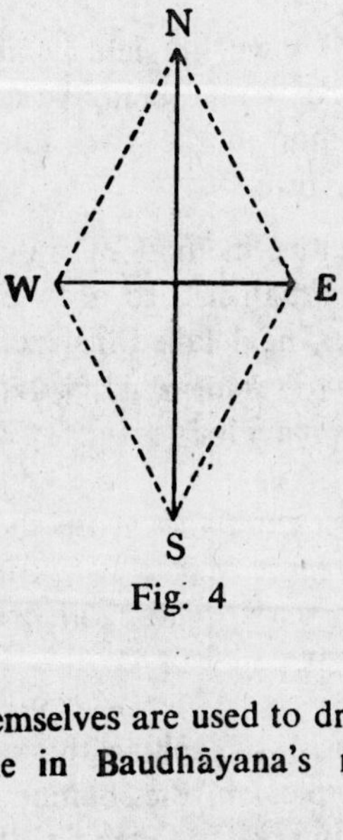

Fig. 4

2.7.2. Such arcs themselves are used to draw two perpendicular diameters in a circle in Baudhâyana's recipe for drawing a square.

—लेखामालिख्य, तस्य मध्ये शंकुं निहन्यात्तस्मिन् पाशौ प्रतिमुच्य लक्षणेन मण्डलं परिलिखेत् विष्कम्भान्तयोः शंकू निहन्यात् ।
पूर्वस्मिन् पाशं प्रतिमुच्य पाशेन मण्डलं परिलिखेत् एवमपरस्मिन् । ते यत्र समेयातां तेन द्वितीयं विष्कम्भमायच्छेत् ।

(*B. Sl.* I. 22-25)

(Drawing a line one fixes a pin at its middle. Slipping the end ties on to this pin, one draws a circle with the mark (the middle mark of the cord) and fixes pins at the ends of the diameter. With the end-tie on the eastern pin one draws a circle with the whole cord. Similarly at the western pin. The second diameter should be stretched through the points where these (circles) intersect.)

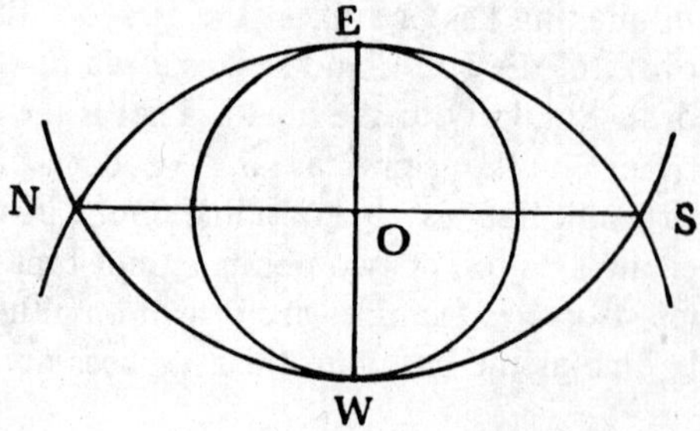

Fig. 5

This method is well-known in later Indian mathematics as the 'fish' method, the name being based on the fact that the lenticular portion common to the two intersecting circles has roughly the shape of a fish.

2.7.3. The most primitive method of drawing a square with a bamboo employs the 'half-fish' to draw a perpendicular at the middle point of a line. The detailed instructions are:

यावान् यजमान ऊर्ध्वबाहुस्तावदन्तराले वेणोश्छिद्रे करोति मध्ये तृतीयम् ।
अपरेण यूपावटदेशं अनुपृष्ठ्यं वेणुं निधाय छिद्रेषु शंकून् निहत्य उन्मुच्यापराभ्यां दक्षिणाप्राक् परिलिखेदान्तात् ।
उन्मुच्य पूर्वस्मादपरस्मिन् प्रतिमुच्य दक्षिणा प्रत्यक् परिलिखेदान्तात् ।
उन्मुच्य वेणुं मध्यमे शंकावन्त्यं वेणोश्छिद्रं प्रतिमुच्य उपर्युपरि लेखासमरं दक्षिणा वेणुं निधाय, अन्त्ये छिद्रे शंकुं निहत्य, तस्मिन् मध्यमं वेणोश्छिद्रं प्रतिमुच्य, लेखान्तयोरितरे प्रतिष्ठाप्य छिद्रयोः शंकुं निहन्ति, स पुरुषश्चतुरश्रः ।

(*Āp. Sl.* VIII. 8-10 & IX. I.)[1]

(At an interval which is as much as the sacrificer with uplifted hands, he makes two holes on the bamboo and a third one at the middle (of this interval). Placing the bamboo along the *pṛṣṭhyā* to the west of the *yūpāvaṭas* (pits for the sacrificial posts) and fixing pins at the holes he releases the bamboo from the western pins and draws (an arc) towards the south-east from the (western) end. Releasing the bamboo from the eastern pin and fixing it on the western pin he draws towards the south-west beginning from the (eastern) end. Releasing the bamboo and fixing the terminal hole of the bamboo on the middle pin (of the

[1]See also *B. Sl.* iii. 13 ff.

pṛṣṭhyā) and placing the bamboo towards the south over the point of intersection of the lines (arcs) he fixes a pin at the end of the bamboo. Then slipping the middle hole on this and adjusting the others at the end of the arcs (i.e. to touch the arcs) he fixes pins at the holes. That is the square of one *puruṣa*). Let AB be the *pṛṣṭhyā* and O its middle point. A bamboo equal to A B in length, is laid along A B pivoted at A and rotated through a right angle so that the free end of the bamboo reaches the south-east corner, thus drawing an arc. Again the bamboo is pivoted at B and the free end drawn from A to the south-west corner tracing another arc. The bamboo is now placed so as to join O to the intersection O′ of the arcs, when it will be perpendicular to AB at O.

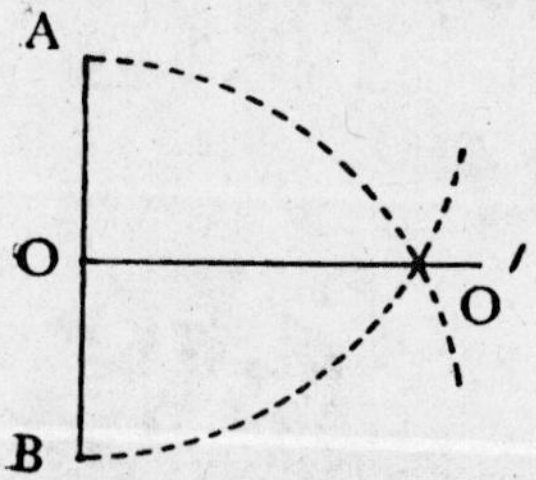

Fig. 6

2.7.4. The method of 2.7.1 for the perpendicular bisector is used for drawing a perpendicular at any given point in one of the methods for the construction of a rectangle given by Baudhāyana as also in one of the methods for drawing a square given by Āpastamba[1] and Manu[2].

दीर्घंचतुरश्रं चिकीर्षंन् यावच्चिकीर्षेत् तावत्यां भूमौ द्वौ शंकू निहन्यात् ।
द्वौ द्वावेकैकमभितः समौ ।
यावती तिर्यङ्मानी तावतीं रज्जुमुभयतःपाशां कृत्वा मध्ये लक्षणं करोति पूर्वेषामन्त्ययोः पाशौ प्रतिमुच्य लक्षणेन दक्षिणापयम्य लक्षणे लक्षणं करोति सोऽसः ।

(*B. Sl.* I. 36-40)

(Wishing to construct a rectangle one should fix two pins at an interval which is as much as is desired. On either side of each at equal distances, two more pins (are fixed). Making ties at the ends of a cord equal in length to the horizontal side, one makes a mark at the middle. Fixing the ties to the terminal ones of the pins at the east and stretching (the cord) by the middle mark to the south one should make a mark (on the ground) at the middle mark. Fixing the ties to the middle pin one stretches (the cord) by the middle mark to the south over the mark already

[1] *Āp. Sl.* I. 7.
[2] *Mā. Sl.* p. 5.

made and marks the middle. This is the top corner.)

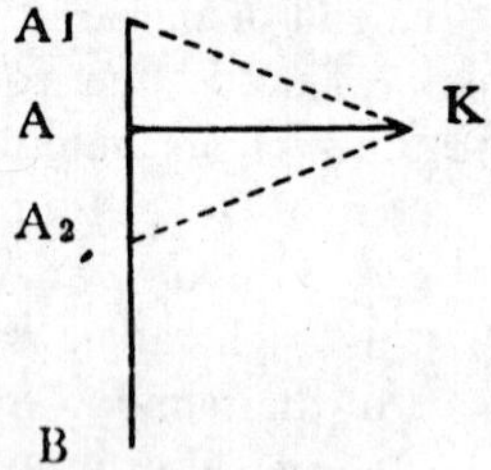

Fig. 7

If perpendiculars are required at A and B on the line AB, two points A_1 and A_2 are marked at equal distances from A. Then an isosceles triangle is drawn on $A_1 A_2$ as base. The vertex K is joined to A. Then A K is the perpendicular at A. The method for drawing the isosceles triangle is to tie ends of a cord to A_1 and A_2 and stretch it tight by its middle point.

2.7.5. The device most frequently adopted for drawing a perpendicular is to use a cord divided into two parts so that the parts form the sides of a right-angled triangle with the line or part of the line on which the perpendicular is to be drawn. One instance will make the procedure clear. For making the *Mahāvedi*, which is an isosceles trapezium of altitude 36 *prakramas*, base 30 *prakramas* and top 24 *prakramas*, Āpastamba directs,

षट्-त्रिंशिकायां अष्ठादशोपसमस्य अपरस्मादन्तात् द्वादशसु लक्षणं, पञ्जदशसु लक्षणं पृष्ठ्यान्तयोरन्तौ नियम्य, पंचदशिकेन दक्षिणापयम्य शंकुं निहन्ति। एवं उत्तरतः। ते श्रोणी...... ।

(V. 2)

(To a cord of 36 *prakramas* 18 *prakramas* are added. Marks are made at distances of 12 and 15 *prakramas* from the other end. Fixing the ends (of the cord) to the end of the *pṛṣṭhyā*, one stretches it to the south by the 15th mark and fixes a pin. Similarly to the north. These are the bottom corners.)

The cord is $36+18=54$ *prakramas* in length. The 12th mark is for measuring half the top side while the 15th mark is both for

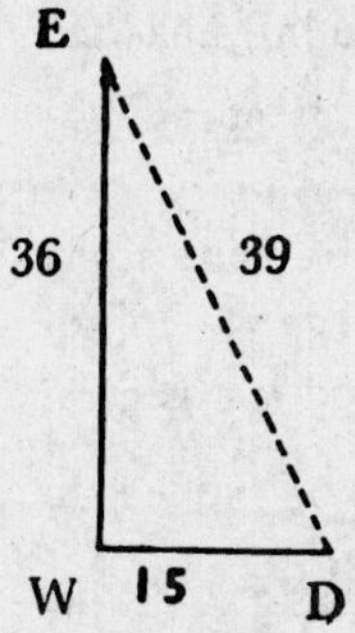

Fig. 9.8

measuring the base and for making the right triangle. For the remainder of the cord is 36+3 = 39 *prakramas*. Therefore the parts of the cord 39 and 15 *prakramas* respectively will make a right triangle with the *pṛṣṭhyā* which is 36 *prakramas*, in length. Hence W D will be at right angles to E W. Āpastamba gives other rational right triangles too which can be used for making a right angle in lieu of the 39,36,15 one. These are (1) 3, 4, 5 which numbers can be multiplied by 4 and 5 to suit the measurements required by the *Mahāvedi* (V. 3) (2) 12,5,13 which can be multiplied by 3 for marking the bottom corners (V. 4), (3) 15, 8, 17 (V. 5), (4) 12, 35, 37, (V. 5). All these are chosen so as to make at least one of the sides have the length of one element of the *Vedi*.

A slight modification makes the method suitable for any length of the given side. To cite from Āpastamba again :

यावदायामं प्रमाणं ।
तदर्धमभ्यस्यापरस्मिंस्तृतीये षड्भागोने लक्षणं करोत्ति ।
पृष्ठ्यान्तयोरन्तौ नियम्य लक्षणेन दक्षिणापयम्य निमित्तं करोति ।
एवमुत्तरतो विपर्यस्येतरतस्स समाधिः ।

(I. 2)

(The measure (of the cord) is the length (of the *prācī*). Adding half of itself to it one makes a mark at the latter third of the cord diminished by one-sixth of that third. Fixing the ends to the ends of the *pṛṣṭhyā* one stretches (the cords) to the south by the mark and makes a mark. Similarly to the north. Then changing to the other side. This is the construction).

By this process, if x is the length of the *pṛṣṭhyā* the cord is $x + \frac{x}{2}$ in length and the mark is at a distance of $\frac{x}{2} - \frac{x}{2.6}$ $= \frac{5x}{12}$. The other part of the cord is $x + \frac{x}{2.6} = \frac{13x}{12}$. Hence when the ends of the cord are tied to the ends of the *pṛṣṭhyā* and the cord is stretched by the mark, we get a triangle whose sides are x, $\frac{5x}{12}$ and $\frac{13x}{12}$ i.e. in the proportion 12, 5 and 13. Therefore it is a right triangle.

The next *sūtra* teaches the use of another right triangle similarly made with sides equal to x, $\frac{3x}{4}$ and $\frac{5x}{4}$. Baudhāyana and Kātyāyana too give these constructions (*B. Sl.* I. 29-35 and I. 42-44; *K. Sl.* I. 12-15)

Āpastamba's commentators Kapardin and Karavinda think that the particle '*vā*' in "आयामं वाभ्यस्य आगन्तुचतुर्थेन ... ..." of *Āp. Sl.* I. 3 is meant to indicate that the method is a general one. Any fraction $\frac{1}{n}$ of the original length can be added to the cord. The original length combined with $\frac{1}{2(n+1)}$ of the added part will be the hypotenuse and the rest of the cord will be the smaller perpendicular side. This would mean Āpastamba here gives a general solution of the rational right triangle with one side given, which is :

$$x, \left(\frac{x}{n} - \frac{x}{2n(n+1)}\right) \text{ and } \left(x + \frac{x}{2n(n+1)}\right)$$

$$\text{i.e. } x, \; x\frac{2n+1}{2n(n+1)} \text{ and } x\frac{2n^2+2n+1}{2n(n+1)}$$

or $x.2n(n+1)$, $x(2n+1)$ and $x(2n^2+2n+1)$

The solution can be verified easily.

$$\text{For, } \{2n(n+1)\}^2 + (2n+1)^2.$$
$$= 4n^2(n^2+2n+1)+4n^2+4n+1$$
$$= 4n^4+8n^3+8n^2+4n+1$$
$$= (2n^2+2n+1)^2$$

A cord divided into x and $x.\sqrt{2}$ is also used (*Āp. Sl.* II. 1)

2.8. *Construction of squares and rectangles*

Most of the above constructions for the perpendicular occur in connection with the construction of squares, rectangles and trapezia. Four distinct methods are detailed for the construction of the square, one of which is applicable to the rectangle also.

2.8.1. A straight bamboo of the length of the side of the square has holes at its ends and middle. With the bamboo lying along the *pṛṣṭhyā* pins are fixed at the three holes. Then by the method described in construction (2.7.3) for the perpendicular, the bamboo is placed at right angles to the *pṛṣṭhyā* at its centre and the end of the bamboo is marked with a pin. Now the middle hole of the bamboo is slipped on to this pin and the bamboo is adjusted so that its ends touch the arcs. The ends of the bamboo then mark the corners of the square. Since the bamboo in its final position is tangential to the two arcs, it must be perpendicular to the radii at the points of contact and these radii are the two horizontal sides of the square. Hence the construction.[1]

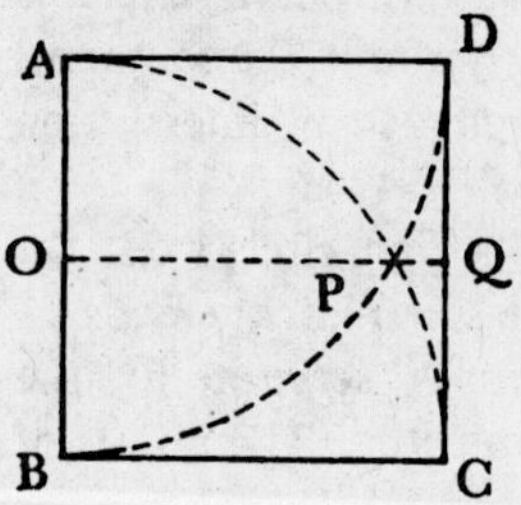

Fig. 9

Dr. B. B. Datta points out that the practice of measuring out the fire-altar with the bamboo-rod is mentioned in the *Taittirīya Samhitā* (V.2.5.1 ff) and in the *Maitrāyanī* (iii. 2.4) *Kaṭha* (xx . 3-4) and *Kapiṣṭhala* (xxxii. 5.6) *Samhitās*. It is perhaps even older than all these *samhitās* since in all these fire is mythically connected with the bamboo.[2]

2.8.2. Baudhāyana's first method for the construction of a square with a given side results in a beautiful geometrical pattern.

चतुरश्रं चिकीर्षन् यावच्चिकीर्षेत् तावतीं रज्जुमुभयतःपाशां कृत्वा मध्ये लक्षणं करोति लेखामालिख्य तस्या मध्ये शंकुं निहन्यात्तस्मिन् पाशौ प्रतिमुच्य लक्षणेन मण्डलं परिलिखेत्,
विष्कम्भान्तयोः शंकू निहन्यात् ।
पूर्वस्मिन् पाशं प्रतिमुच्य पाशेन मण्डलं परिलिखेत् । एवमपरस्मिन् । ते यत्र समेयातां तेन द्वितीयं विष्कम्भमायच्छेत् ।
विष्कम्भान्तयोः शंकू निहन्यात् ।
पूर्वस्मिन् पाशौ प्रतिमुच्य लक्षणेन मण्डलं परिलिखेत् एवं दक्षिणात् एवं पश्चादेवमुत्तरतस्तेषां येऽन्त्याः संसर्गास्तच्चतुरश्रं सम्पद्यते ।

(*B. Sl.* I. 22-28)

(Wishing to construct a square one should make ties at both ends of a string as long as the desired side and make a mark at its middle. One should draw a line and fix a pin at its middle. Fixing the ties on this pin one should draw a circle by the

[1] *Āp. Sl.* VIII. 8-10 and IX.—1 and *B. Sl.* III. 13 ff. Āpastamba's *sūtras* have already been quoted.

[2] Datta, *Science of the Śulba*, pp. 60-61 and the footnotes thereto.

middle mark (of the cord) and at the ends of the diameter (formed by the *prācī*) one should fix pins. Fixing one tie on the eastern pin one should draw a circle with the other tie. Similarly round the western pin. Through the points where they meet the second diameter should be drawn and pins should be fixed at its ends. With the ties on the eastern pin a circle is to be drawn with the middle mark. Similarly round the southern, western and northern (pins). Their outer points of intersection form the square).

The explanation of the procedure is: The first circle fixes the middle points of the horizontal sides of the square or the ends of the line of symmetry. The two bigger circles serve to fix the perpendicular bisector of the line of symmetry. The points of

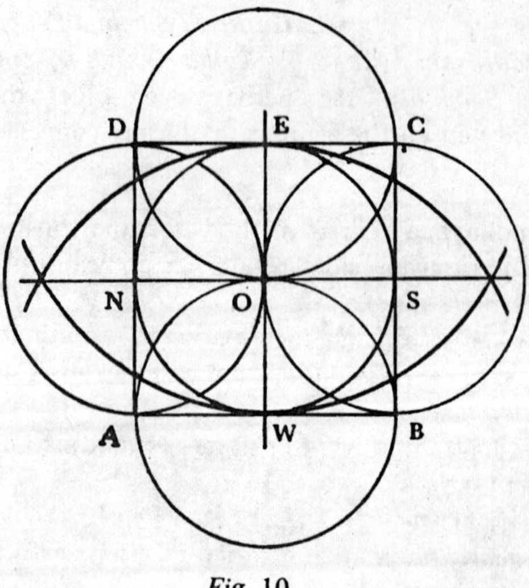

Fig 10

intersection of this with the first circle are the middle points of the vertical sides. Hence the vertices of the square, which will be equidistant from pairs of the middle points of the sides, will be obtained as the points of intersection of circles drawn with the middle points as centres and half the side as radius, taken two by two. This method is confined to the *Baudhāyana Śulba-sūtra*.

2.8.3. The first construction for a square given by Āpastamba is:

प्रमाणमात्रीं रज्जुमुभयत:पाशां करोति । मध्ये लक्षणमर्धमध्यमयोश्च । पृष्ठ्यायां रज्जुमायम्य पार्श्वयोर्लक्षणेषु इति शंकुं निहत्य उपान्त्ययो: पाशौ प्रतिमुच्य मध्यमेन लक्षणेन दक्षिणापयम्य शंकुं निमित्तं करोति । मध्यमे पाशौ प्रतिमुच्य उपर्युपरि निमित्तं मध्यमेन लक्षणेन दक्षिणा-पयम्य शंकुं निहन्ति । तस्मिन् पाशं प्रतिमुच्य पूर्वास्मिन्नितरं मध्यमेन लक्षणेन दक्षिणमंसं श्रायच्छेत् । उन्मुच्य पूर्वस्मात् श्रपरस्मिन् प्रतिमुच्य मध्यमेनैव लक्षणेन दक्षिणां श्रोणिमायच्छेत् । एवमुत्तरौ श्रोण्यंसौ ।

(I. 7)

(A cord of the length of the *pṛṣṭhyā* is noosed at either end. Marks (are made) at the middle and the middles of the halves. Stretching the cord along the *pṛṣṭhyā* one stands pins at the nooses (P and Q) and the marks (R, O, S). Inserting the nooses on the pins next to the terminal ones (i.e. R and S) and stretching the cord southwards by the middle mark one makes a mark on the ground. Again inserting both the nooses on the middle pin (O) one stretches the cord by the middle mark southward over and beyond the mark on the ground and fixes a pin (E). Inserting one noose on this pin and the other on the eastern pin (P) one should stretch the cord by the middle mark for the south top corner (C). Taking off from the eastern (pin) one fixes the noose on the western pin (Q) and stretches the cord by the middle mark for the south bottom corner (B). Similarly the top and bottom corners of the north.

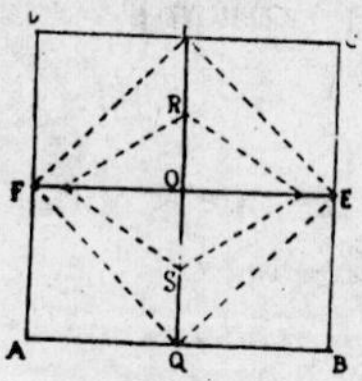

Fig. 11

The principles underlying this construction are the same as those of the previous construction. But cord-stretching, which leaves the final drawing free of all unnecessary lines, is resorted to in place of the drawing of circles.

This method is given by Manu[1] and Kātyāyana[2] also.

[1]*M. Sl.* I. p. 5.

[2]*Kāt. Śr.* XVI. 8. 1-20 as quoted by B. Datta, *Science of the Śulba*, p. 62.

2.8.4. The next method, equally useful for the square and the rectangle, is to draw perpendiculars at the ends of the *pṛṣṭhyā* and then mark off the length of the side on these perpendiculars to get the corners. For drawing the perpendicular, the method of drawing the isosceles triangle with a cord can be used as in Baudhāyana's method for the construction of a rectangle (I. 36-40). Or the cord can be used to form a right triangle with the *pṛṣṭhyā* as in Baudhāyana's construction for the square (*B. Sl.* I. 29-35) and Āpastamba's construction for any rectangular figure (*Āp. Sl.* I. 2 and 3). Āpastamba achieves the construction of the square with one operation the less, by constructing right triangles on half the *pṛṣṭhyā* with a cord divided into two parts respectively equal to half the side and $\sqrt{2}\,\frac{\text{side}}{2}$. The direction is:

पृष्ठ्यान्तयोर्मध्ये च शंकुं निहत्यार्धे तद्विशेषमभ्यस्य लक्षणं कृत्वा अर्धमागमयेत् । अन्तयोः पाशौ कृत्वा, मध्यमे सविशेषं प्रतिमुच्य, पूर्वस्मिन्नितरं, लक्षणेन दक्षिणमंसमायच्छेत् । उन्मुच्य पूर्वस्मादपरस्मिन् प्रतिमुच्य, लक्षणेनैव दक्षिणां श्रोणिमायच्छेत् । एवमुत्तरौ श्रोण्यंसौ ।

(*Āp. Sl.* II. 2)

(Fixing pins at the ends and middle of the *pṛṣṭhyā* one should add to a cord of half the length its *viśeṣa* ($\sqrt{2}-1$ of any length is known as its *viśeṣa* in the *Śulbasūtras*), make a mark there *and* then add half the side again. Making nooses at the ends, and fixing the side with the *viśeṣa* to the middle pin, and the other side to the eastern pin, one should stretch (the cord) by the mark for the southern top corner. Taking off from the eastern pin and inserting on the western pin one should stretch (the cord) by the same mark for the south bottom corner. Similarly the northern top and bottom corners.) (Fig. 12). In IX. 3 Āpastamba uses two bamboo rods equal to one *puruṣa* and $\sqrt{2}$ *puruṣas* respectively for constructing a square one sq. *puruṣa* in area.

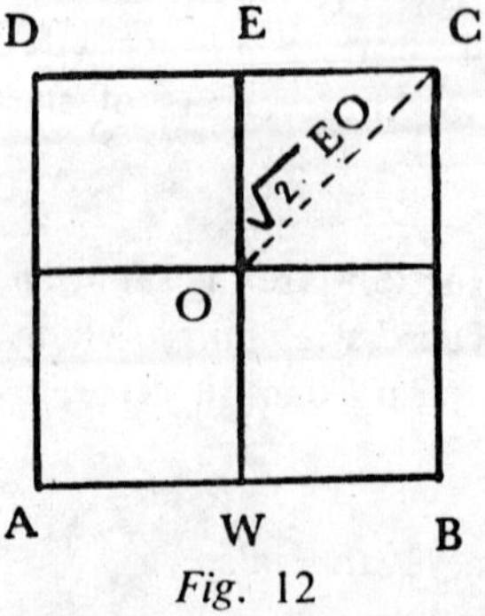

Fig. 12

2.9. *Construction of a trapezium with the face, base and altitude given*

The name for a trapezium used by the *Śulbasūtras* is *ekato'ṇimat* (smaller on one side).[1] *Mahāvedi*, the *vedi* for the *soma* sacrifice of paramount importance and other *vedis* too were to be trapezia in shape. Hence the *Śulbasūtras* dwell on its construction in detail. The method used is essentially the same as for the construction of the square and the rectangle using a right-angled triangle. But various rational right triangles which will suit the measurements of the *Mahāvedi* are given. Once the perpendiculars at the ends of the *pṛṣṭhyā* (the line of symmetry or altitude) are drawn, the lengths of the top and bottom sides can be marked off on them. Since the measurements of the *Mahāvedi* are altitude = 36 *prakramas*, base = 30 *prakramas* and face = 24 *prakramas*, the eligible right triangles, according to Āpastamba, are:

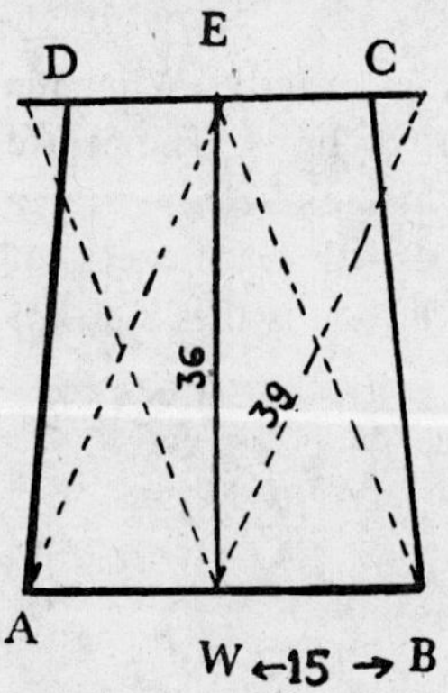

Fig. 13

36, 15, 39	(*Āp. Śl.* V. 2)
3, 4, 5 multiplied by 4 & 5	(*Āp. Śl.* V. 3)
12, 5, 13	(*Āp. Śl.* V. 4)
15, 8, 17	(,, ,, V. 5)
12, 35, 37	(,, ,, ,,)

Baudhāyana recommends his method for constructing a rectangle for the construction of a trapezium also (I. 36-41) i.e. drawing the perpendiculars at the extremities of the *pṛṣṭhyā* by the isosceles triangle method and then marking off the half sides on these.

2.10. *Transformation of Figures*

The votive fire-altars were prescribed different shapes according to the specific benefit sought for — the *śyenacit* (the fire-place in the form of a falcon) for attaining heaven, the *praugacit* (the fire-place in the form of an isosceles triangle) for destroy-

[1] *B. Śl.* I. 72.

ing enemies and so on. But all these different shapes had to have strictly the same area, viz. $7\frac{1}{2}$ sq. *puruṣas*. Hence were evolved methods for transforming one geometrical figure into another, more especially the square into other equivalent geometrical figures. These constructions are given below.

2.10.1. *To convert a square into a circle*

No geometrical method can achieve this exactly. What the *Śulbasūtras* do is to give approximate constructions. The centre O of the square is joined to a vertex A and the circle is drawn with half the side of the square combined with $\frac{1}{3}$ the excess of OA over half the side of the square,[1] i.e. if '*a*' is the side of the square and '*r*' the radius of the circle.

$$r = \frac{a}{2} + \frac{\sqrt{2}.\frac{a}{2} - \frac{a}{2}}{3}$$

$$= \frac{a}{2} \times \frac{(2+\sqrt{2})}{3}$$

The value of π calculated from this is only about 3.088. But according to some of the commentators, the last sentence of this rule, namely *Sānityā maṇḍalam*, is to be split as *Sā anityā maṇḍalam*, when it will mean that Āpastamba and the other authors of *Śulbasūtras* as well were aware that this was an approximate method only. Thibaut and Bürk, understandably, do not accept this explanation.

2.10.2. *To convert a circle into a square*

All the three important *Śulbas* direct us to divide the diameter into 15 parts and to take 13 of these parts as the side of the square i.e., if d is the diameter of the circle[2] and a the side of the equivalent square,

$$a = \frac{13}{15}d, \text{ whence } \pi = 3.004$$

Baudhāyana gives a slightly better approximation too.

मण्डलं चतुरश्रं चिकीर्षन् विष्कम्भमष्टौ भागान् कृत्वा भागमेकोनत्रिंशता विभज्य अष्टाविंशति भागानुद्धरेत् भागस्य च षष्ठं अष्टमभागोनम् । (*B. Sl.* I. 59)

[1] *Āp. Sl.* III. 2; *B. Sl.* I. 58 and *K. Sl.* III.
[2] *Āp. Sl.* III. 3; *B. Sl.* I. 60 and *K. Sl.* 14

(Wishing to convert a circle into a square one should divide the diameter into 8 parts, divide one of these parts again into 29 parts and subtract 28 of these (29th parts) together with one-sixth of one of these parts, the latter being diminished by one-eighth of that (one-sixth part)).

$$\text{i.e., } a = d\left(1 - \frac{28}{8.29} - \frac{1}{6.8.29} + \frac{1}{6.8.29.8}\right)$$

This value is based on an inversion of the relation between *r* and *a* given in connection with the problem of circling the square. How exactly the value was brought to the form of this long and complicated fractional expression is a matter for speculation, but may not be of geometrical interest.

2.10.3. *To convert a rectangle into a square*

Āpastamba's rule is:

दीर्घचतुरश्रं समचतुरश्रं चिकीर्षन् तिर्यङ्मान्यापच्छिद्य शेषं विभज्योभयत उपदध्यात्। खण्डमागन्तुना संपूरयेत्। तस्य निह्रास उक्तः।

(*Āp. Sl.* II. 7)

(Wishing to turn a rectangle into a square, one should cut off a part equal to the transverse side and the remainder should be divided into two and juxtaposed at the two sides (of the first segment). The bit (at the corner) should be filled in by an imported bit. The removal of this has been explained already).

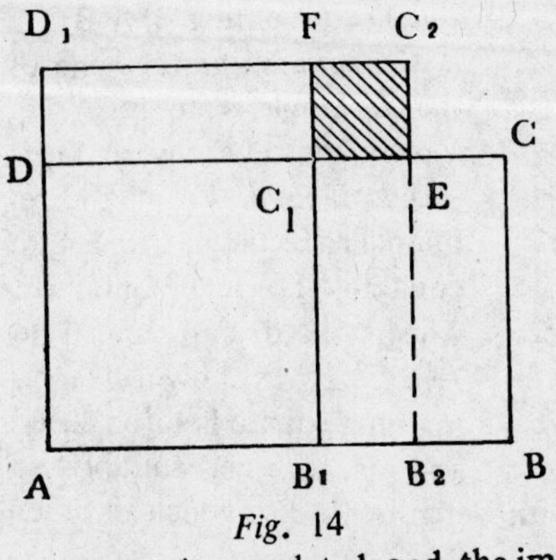

Fig. 14

If ABCD is the rectangle, a square AB_1C_1D with side = AD, is cut off from it. The remaining rectangle B_1BCC_1 is divided into two equal strips $B_1B_2EC_1$ and B_2BCE. The strip B_2BCE is cut off and applied to the side of the square DC_1. Now we get a square of side AD_1 with a small square C_1EC_2F unfilled up at one corner. The larger square is completed and the imported square in the corner is removed by the method of removing a square from a square, (given in 2. 11. 4.).

Baudhāyana (I. 54) and Kātyāyana (III. 2) give the same method. Though this method works with any rectangle,[1] Kātyāyana provides for a very long rectangle with a separate *sūtra*.

अतिदीर्घं चेत्तिर्यङ्मान्यौपच्छिद्य अपछिद्यैकसमासेन समस्य शेषं यथायोगमुपसंहरेत् ।

(*K. Sl.* III. 3)

(If the rectangle) is very long, cut off repeatedly by the transverse side (breadth) and join the squares so formed into one big square, and then the remainder of the rectangle should be joined to this square as it fits (to form a square).

The method is no improvement over the general method, since here no side of the remainder rectangle will be equal to the side of the bigger square to which its strips are to be joined.

2.10.4. *To convert a square into a rectangle*

समचतुरश्रं दीर्घचतुरश्रं चिकीर्षन् यावच्चिकीर्षेत् तावतीं पार्श्वमानीं कृत्वा यदधिकं स्यात् यथायोगमुपदध्यात् ।

(*Āp. Sl.* III. 1)[2]

(Wishing to convert a square into a rectangle one should make the lateral side as long as is desired and the excess should be joined suitably.)

We are not told how exactly the excess is to be joined. Thibaut and Bürk suggest that this was achieved by repeated slicing and joining. If A B C D is the square and the side of the rectangle is to be A D_1, a rectangle A B by A D_1 is sliced off first. From the remainder a rectangle with length equal to A D_1 can be obtained. This is sliced and joined to A B C_1 D_1 as shown. The remaining square is to be sliced and put together suitably so as to get a narrow strip with length = A D_1, which is then

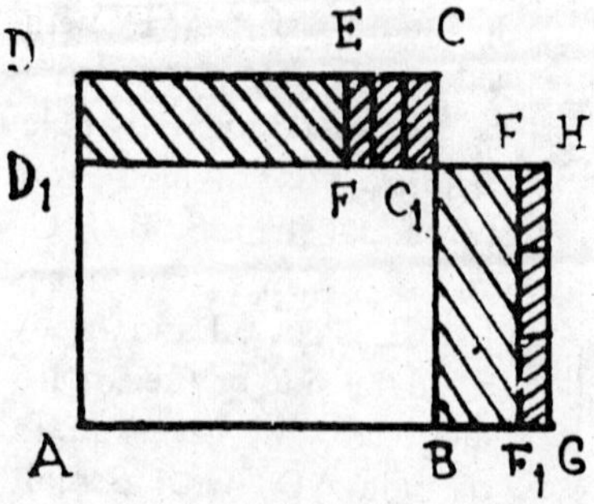

Fig. 15

[1]If the length is greater than thrice the breadth, the *āgantu* or imported square will be the bigger square.

[2]Also *B. Sl.* I. 53.

again joined to A F_1 F D_1. This procedure is not merely empirical but also highly unsatisfactory, since there is no guarantee that the remainder square always yields a rectangle of the requisite length.

Āpastamba's commentator Sundararāja, explaining this *sūtra*, gives a purely geometrical and exact construction.

यावदिच्छं पार्श्वमान्यौ प्राच्यौ वर्धयित्वा उत्तरपूर्वां कर्णरज्जुमायच्छेत्, सा दीर्घचतुरश्रमध्यस्थायां समचतुरश्रतिर्यङ्मान्यां यत्र निपतति तत उत्तरं हित्वा दक्षिणांशं तिर्यङ्मानीं कुर्यात्, तद् दीर्घचतुरश्रं भवति ।

(Producing the sides of the square eastward to the desired length of the lateral side, one should draw the north-eastern diagonal. The part of the transverse side to the north of the point where the diagonal cuts it is to be discarded and its southern part is to be made the transverse side of the rectangle. That will be the rectangle.)

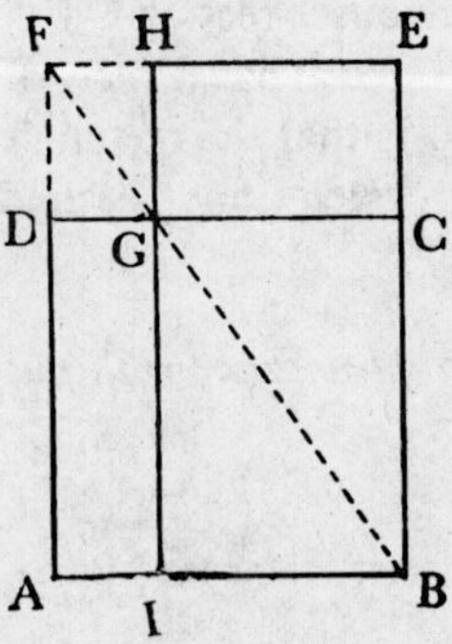

Fig. 16

Let A B C D be the given square. Produce A D and B C to F and E so that A F = B E = the required side of the rectangle. Complete the rectangle A B E F and join the diagonal B F cutting C D in G. Through G draw a st. line IH parallel to the sides of the square. Then IBEH is the required rectangle.

For, F A B = F E B
I G B = G C B
and F D G = F H G

Hence rect. A I G D = rect. G C E H

∴ rect. I B E H = rect. I B C G + rect. G C E H
= rect. I B C G + rect. A I G D
= sq. A B C D

Here the case where the given side of the rectangle is greater than the side of the square only is dealt with. But with slight change the method is applicable to the case where the side is less also.

Baudhāyana and Kātyāyana give an easier method.

समचतुरश्रं दीर्घचतुरश्रं चिकीर्षन् मध्येऽक्षणयापच्छिद्य विभज्येतरत्पुरस्तादुत्तरतश्चोपदध्यात् । विषमं चेद्यथायोगं उपसंहरेदिति व्यासः । (*K. Sl.* III. 4)[1]

[1] Also *B. Sl.* I. 52.

(Wishing to transform a square into a rectangle one should cut diagonally in the middle, divide one part again and place the two halves to the north and east of the other part. If the figure is a quadrilateral one should place together as it fits. This is the distribution). Let A B C D be the square. It is cut along AC to form two right triangles. The △ A B C is again halved along the altitude B E. The two halves B E A and B E C are then removed to the positions D F C and D G A.

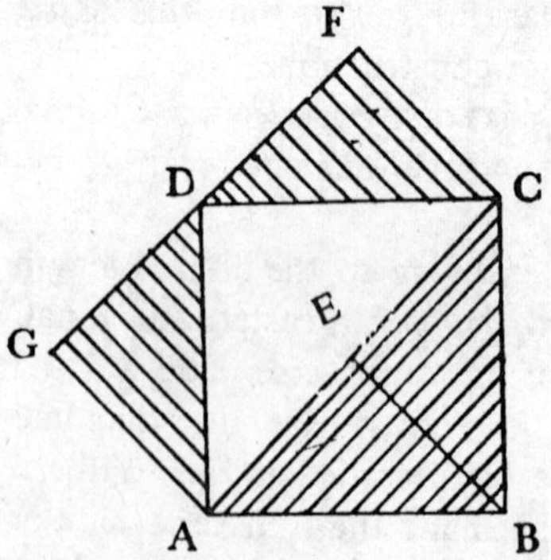

Fig. 17

Then, obviously rect. A C F G = sq. A B C D. The defect of this method is that the rectangle cannot be given any desired side.

2.10.5. *To convert a rectangle or square into a trapezium with the shorter parallel side given.*

Baudhāyana deals with this problem.

चतुरश्रमेकतोऽणिमच्चिकीर्षन् अणिमतः करणीं तिर्यङ्मानीं कृत्वा शेषमक्ष्णया विभज्य विपर्यस्येतरत्रोपदध्यात् ।

(*B. Sl.* I. 55)

(If one wishes to make a square or rectangle shorter on one side, one should cut off a portion by the shorter side. The remainder should be divided by the diagonal, inverted and attached on either side).

If A B C D is the given rectangle, let the rectangle A F E D be cut off so that A F = D E = the given shorter side. The remaining rectangle E F B C is to be cut diagonally along B E and the portion B E C is to be inverted and attached to the side A D of the rectangle in the position E′ A D. Then D E′ B E is the equivalent trapezium.

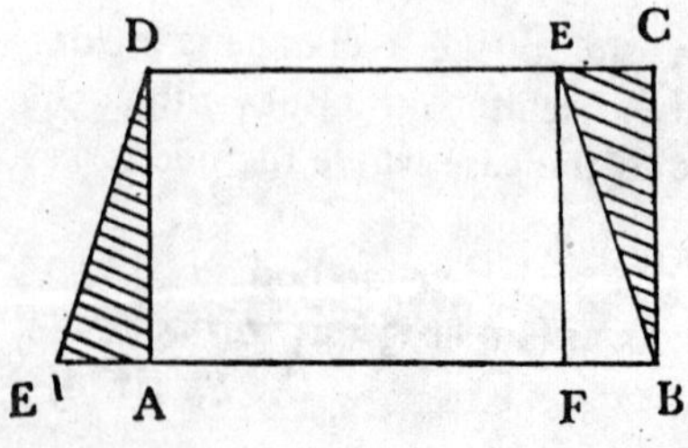

Fig. 18

The *Śatapatha Brāhmaṇa* seems to give another method for this conversion.[1]

चतुर्विंशत्यंगुलिभिर्मिमीते । चतुर्विंशत्यक्षरा वै गायत्री, गायत्रोऽग्निर्यावानग्निर्यावत्यस्य मात्रा तावतैवैनं तान्मिमीते । स चतुरंगुलमेव उभयतोऽन्तरत उपसमूहति, तावद् व्यदूहति । तन्नाहेवातिरेचयति नो कनीयः करोति ।

(He measures by 24 *aṅgulis*. For *Gāyatrī* has 24 letters and *agni* is of *Gāyatrī*. As much as is the *agni*, as much as is its measure, by that much he measures them. He contracts 4 *angulis* inwards on both sides. He stretches 4 *angulis* outwards on both sides. as much as he contracts, so much he stretches. Thus he does not make it exceed (the right measure) nor does he make it smaller).

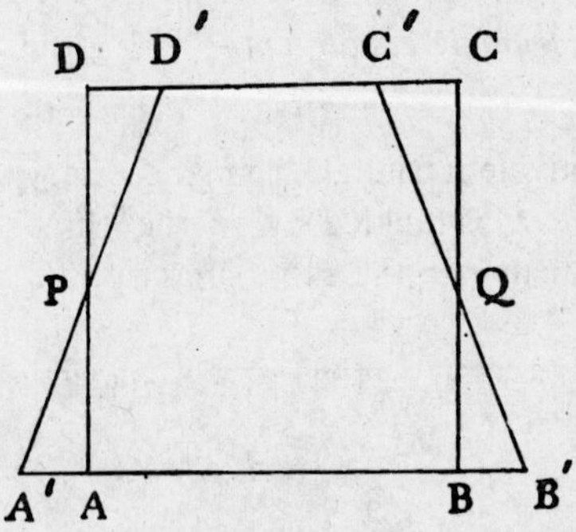

Fig. 19

Datta takes this to mean[2] that the face of the square (A B C D) is shortened on either side by a small length (D D′ = C C′) and the base is lengthened on either side by the same length to A′ and B′. Then the area of trapezium A′B′C′D′ = the area of square ABCD. and obviously so, since

$\triangle PDD' = \triangle PAA'$

and $\triangle QCC' \equiv \triangle QBB'$

The other *Śulbasūtras* do not deal with this problem.[3]

2.10.6. *To convert a trapezium into an equivalent rectangle*

Āpastamba tackles the converse problem of converting an isosceles trapezium into an equivalent rectangle. It is not given as a general prescription but rather as a means of finding out the area of the trapezium of the Mahāvedi.

दक्षिणस्मादंसाद् द्वादशसु दक्षिणस्यां श्रोण्यां निपातयेत्, छेदं विपर्यस्य उत्तरत उपदध्यात् । सा दीर्घा चतुरस्रा । तथायुक्तां संचक्षीत ।

(From the southern top corner one should drop a perpendicular on the southern bottom corner at a distance of 12 (*padas* from

[1] *Ś. Br.* X 2. 1. 3-4.

[2] *Science of the Śulba*, p. 91.

[3] B.B. Datta says (*Science of the Śulba*, p. 92) that *Āp. Sl.* XV-9 ff deal with this problem, but in the edition I am using the said *sūtras* do not deal with this problem.

the *pṛṣṭhyā*). The removed bit should be placed inverted at the northern side. That is the rectangle. One should study it thus joined).

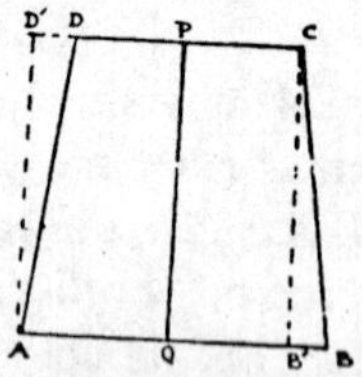

Fig. 20

That is, to convert the trapezium A B C D (parallel sides 24 ft. and 30 ft.), C B′ is drawn perpendicular to A B, B′ being 12 padas away from PQ, the *pṛṣṭhyā*. The triangle C B′ B is then placed in the position A D D′. Then rect. A B′ C D′ = trap. A B C D

2.10.7. *To construct an isosceles triangle equal in area to a given square and vice versa.*

Conversion of a square into an equivalent triangle, being necessary for the construction of the *Praugacit*, is tackled by all the three important *Śulbasūtras* and all of them give the same prescription.

यावानग्निस्सारत्निप्रादेशो द्विस्तावतीं भूमिं चतुरश्रं कृत्वा, पूर्वस्याः करण्या अर्धात् श्रोणी प्रत्यालिखेत् । सा नित्या प्रउगम् ।

(*Ā. Sl.* XII. 5)[1]

(Making an area which is double as much as the fire-altar with the *aratnis* and *prādeśas*, into a square, one should draw lines from the middle point of the eastern side towards the bottom corners. That is the equivalent *prauga* (isosceles triangle).

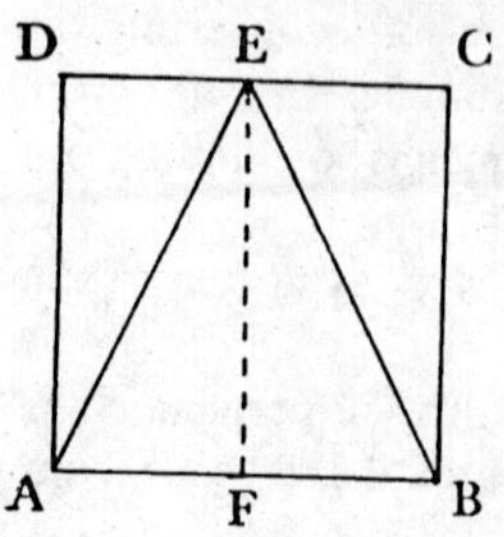

Fig. 21

Let A B C D be a square of twice the required area. Let E be the middle point of C D. E A and E B are joined. Then A E B is the required triangle. For if the altitude EF is drawn, the square is divided into 2 equal rectangles A F E D and F B C E.

A F E = ½ rect. A F E D

and F B E = ½ rect. F B C E

∴ Whole A E B = ½ sq. A B C E.

[1]See also *B. Sl.* I. 56 and *K. Sl.* IV. 5.

This construction leads to the formula,
area of a triangle = $\frac{1}{2}$ base × altitude.

The converse problem is treated by Kātyāyana only.

प्रउगं चतुरश्रं चिकीर्षन् मध्ये प्राच्छामपच्छिद्य विपर्यस्येतरत उपधाय दीर्घचतुरश्रसमासेन समस्येत् । (IV.7)

(Wishing to convert an isosceles triangle into a square, one should cut in the middle towards the east, place on the other side inverted and manipulate by the method of conversion of a rectangle into a square.)

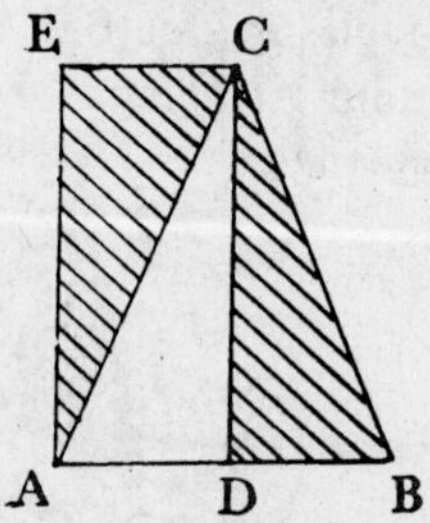

Fig. 22

Let A B C be the triangle. Let it be cut into two halves along the altitude C D. Let the part D B C be applied inverted to the side A C. Then rect. A D C E = △ A B C. Now the rectangle can be converted into an equivalent square by the method already given.

2.10.8. *To construct a rhombus of given area*

तावदेव दीर्घचतुरश्रं विहृत्य पूर्वापरयोः करण्योरर्धात् तावति दक्षिणोत्तरयोर्निपातयेत् सनित्योभयतःप्रउगम् । (*Āp. Sl.* XII. 9)[1]

(Drawing a rectangle of the same area (i.e. of the area of the square for the *prauga*), one should draw lines from the middle points of the eastern and western sides to the middles of the southern and northern sides. That is the rhombus of the same area).

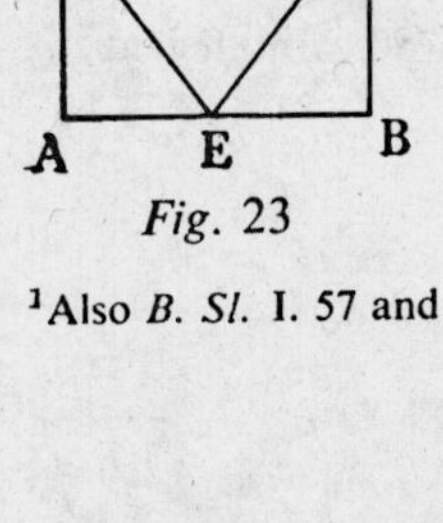

Fig. 23

Let A B C D be a rectangle of twice the area of the rhombus. Let E, F, G, H be the middle points of the sides. E F, F G, G H, H E are joined. Then rhombus E F G H = $\frac{1}{2}$ rect. A B C D.

[1]Also *B. Sl.* I. 57 and *K. Sl.* IV. 6.

2.10.9. *To transform a rhombus into a rectangle*

This converse construction occurs in Kātyāyana only.

उभयतःप्रउगं चेन्मध्ये तिर्यगपच्छिद्य पूर्ववत् समस्येत् ।

(*K. Sl.* iv. 8)

(If it is an *ubhayataḥ prauga* one should cut transversely along the middle and join together as before.)

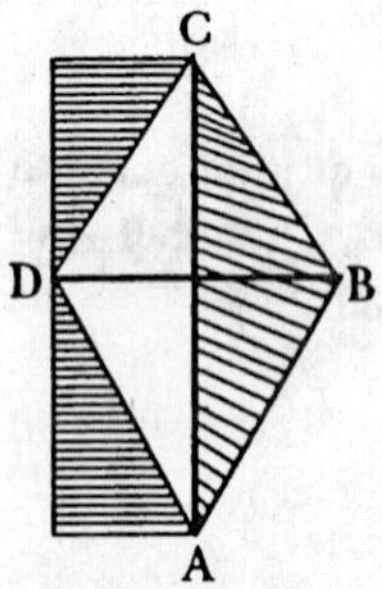

Fig. 25

The process is exactly the same as for the *prauga*. The rhombus is first divided into two isosceles triangles by joining a diagonal and again into 4 right triangles by cutting along their altitudes. The four triangles are joined together to form a rectangle.

2.11.1. *Combination of areas and the converse*

By the application of the theorem of the square of the diagonal the authors of the *Śulbasūtras* combine any number of squares to form another square.

For, combining two equal squares i.e. for doubling a square Āpastamba's rule is:

समस्य द्विकरणी । (I. 5)

(The diagonal of the square is the double-maker.) Hence if a square is drawn on the diagonal of the given square, it will produce double the area. The diagonal will therefore be $\sqrt{2}\,a$ where a is the side of the square. It is noteworthy that the *Śulbasūtras* give a very close approximation to the value of $\sqrt{2}$

प्रमाणं तृतीयेन वर्धयेत्तच्चतुर्थेन भ्रात्मचतुस्त्रिंशोनेन सविशेषः ।

(*Āp. Sl.* I. 5 and *B. Sl.* I. 61-62)[1]

(The measure should be increased by one-third of itself, which again is increased by its one-fourth and diminished by $\frac{1}{34}$ th of that (second) increment. This is the *saviśeṣa*.)

[1]Also *K. Sl.* II. 13.

$$\text{i.e. } \sqrt{2} = 1 + \frac{1}{3} + \frac{1}{3.4} - \frac{1}{3.4.34}$$

(This result must have been arrived at by Rule of Three and the method of repeated correction.[1] For trebling a square

प्रमाणं तिर्यक् द्विकरण्यायामः तस्याक्ष्णयारज्जुस्त्रिकरणी । (*Āp. Sl.* II. 2)

(The breadth is the measure (of the side of the given square) and the length is the double-maker. The diagonal (of such a rectangle) will be the treble-maker.)

In this way proceeding step by step one can combine any number of equal squares.

2.11.1b For combining a large number of squares, Kātyāyana gives an ingenious method in one step.

यावत्प्रमाणानि समचतुरश्राण्येकीकर्तुं चिकीर्षेत् एकोनानि तानि भवन्ति तिर्यक् द्विगुणान्येकत एकाधिकानि । त्र्यस्रिर्भवति तस्येषुस्तत्करोति ।

(*K. Sl.* VI. 7)

The verse is not easy to interpret. The only logical meaning assignable is what Dr. B. B. Datta gives[2], viz. "As many squares (of equal side) as you wish to combine into one, the transverse line will be (equal to) one less than that; twice a side will be (equal to) one more than that. It will be a triangle. Its arrow (i.e. altitude) will do that". That is, if n squares of side a are to be combined, we have to construct an isosceles triangle A B C with (n—1) a as base and $\frac{(n+1)a}{2}$ as sides. AD the altitude is drawn. Then AD is the side of the square whose area will be n a^2.

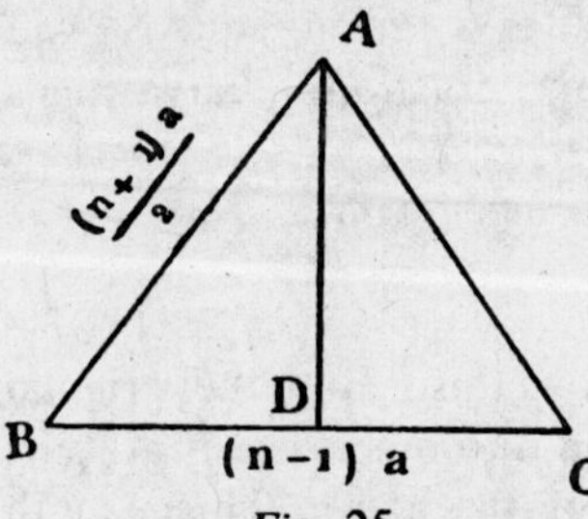

Fig. 25

$$\text{For } BD = \tfrac{1}{2} BC = \frac{(n-1)a}{2}$$

And from the rt-angled $\triangle$ ADB

[1]Thibaut's hypothesis about the deduction of this expression (*J.A.S.B.*, 1875) is ingenious, but one fails to understand the preference for $12^2 + 12^2 = 288 = 17^2 - 1$ when the earlier $2^2 + 2^2 = 8 = 3^2 - 1$ is equally suitable.

[2]*Science of the Śulba*. pp. 72-73.

$$AD^2 = AB^2 - BD^2$$
$$= \left\{\frac{(n+1)a}{2}\right\}^2 - \left\{\frac{(n-1)a}{2}\right\}^2$$
$$= \frac{a^2}{4}\left\{(n+1)^2-(n-1)^2\right\}$$
$$= \frac{a^2}{4} \times 4n = n.a^2$$

Where the number is expressible as the sum of two squares the first method itself can be shortened.

e.g.: $10 = 3^2+1$
$40 = 6^2+2^2$

In such cases we can construct a rectangle with sides 3 a and a or 6a and 2a and then the diagonal will be the side of the combined square.[1]

Or, in general, if $n = p^2+q^2$, one has to construct a rectangle of sides pa and qa. Then the square on the diagonal will be

$$p^2 a^2+q^2 a^2$$
$$= a^2 (p^2+q^2) = n a^2.$$

2.11.2. Methods are also given for getting squares which are fractions of a given square. Since the *Sautrāmaṇī vedi* is to be $\frac{1}{3}$ of the *Saumikī*, the Śulbasūtras deal in detail with the construction of a square whose area is $\frac{1}{3}$ that of a given square. And the method can be extended to any fraction. Kātyāyana's instructions are clearest.

तृतीयकरण्येतेन व्याख्याता । प्रमाणविभागस्तु नवधा करणीतृतीयं नवभागः ।
नवभागास्त्रयः तृतीयकरणी ।

(*K. Sl.* 15-18)[2]

(The one-third-maker is explained by this. But the original measure (of the area) is to be divided into nine. The third part of the side (produces) a ninth part (of the area). Three ninth parts will give the one-third-maker.)

Here we are directed to divide the square into 9 equal parts by dividing the pair of opposite sides into 3 equal parts by lines parallel to the other pair of sides, 3 of the small squares so formed are to be combined into a square, the side of which will then be the one-third-maker.

[1] *K. Sl.* II. 8-9.
[2] *Āp. Sl.* II. 3 and *B. Sl.* I. 47.

The commentators give an alternative explanation also. The triple square is first to be obtained, which is then to be divided into 9 equal parts as above. These parts will be $\frac{1}{3}$ of the original square.

2.11.3. *To combine two unequal squares*

ह्रसीयसः करण्या वर्षीयसो वृध्द्रमुल्लिखेत् । वृध्द्रस्याक्षणयारज्जुरुभे समस्यति ।

(*Āp. Sl.* II. 4)[1]

(With the side of the smaller one a segment of the bigger one should be cut off. The diagonal cord of the segment will combine the two squares).

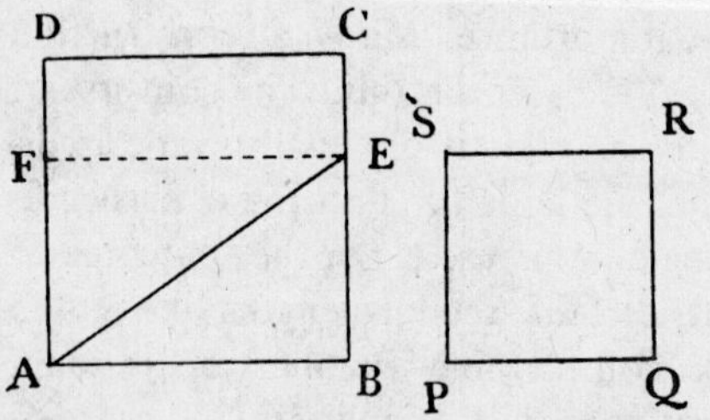

Fig. 26

In effect, a rectangle with sides equal to the sides of the squares is constructed. If a and b are these sides, the square on the diagonal of the rectangular segment $= a^2 + b^2$.

2.11.4. *To draw a square equal to the difference of two squares*

चतुरश्राच्चतुरश्रं निर्जिहीर्षन् यावन्निर्जिहीर्षेत् तस्य करण्या वृध्द्रमुल्लिखेत् ।
वृध्द्रस्य पार्श्वमानीं अक्षणया इतरत् पार्श्वं उपसंहरेत् सा यत्र निपतेत्तदपछिन्द्यात् ।

(*Āp. Sl.* II. 5)[2]

(Wishing to deduct a square from a square one should cut off a segment by the side of the square to be removed. One of the lateral sides of the segment is drawn diagonally across to touch the other lateral side. The portion of the side beyond this point should be cut off.)

[1]Also *B. Sl.* I. 52 and *K. Sl.* II. 22.
[2]Also *B. Sl.* I. 51 and *K. Sl.* III. 1.

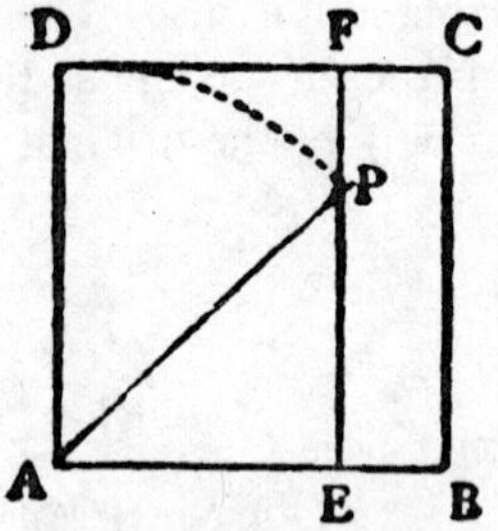

Fig. 27

Let A B C D be the larger square, and A E the side of the square to be removed. The segment A E F D is cut off. A D is drawn diagonally across with A fixed, till D touches E F at P. Then E P is the side of the required square. For, from the right triangle A E P,

$$EP^2 = AP^2 - AE^2 = AD^2 - AE^2.$$

This explanation is given by Āpastamba himself in the next *sūtra*.

2.12. *Geometrical truths implied in the constructions*

The theorem of the square of the diagonal which is also explicitly stated, is indeed, of supreme importance. Many a construction (e.g. the combination of areas) is based on this. The converse of the theorem, though not stated as such, is of equally wide application. One of the commonest methods for drawing a perpendicular is to use the right triangle. This use is not the outcome of a chance observation that certain sets of lengths like 3, 4, 5 produce a right triangle. A considerable number of rational right triangles and a few irrational right triangles[1] with approximate values assigned to the irrational sides are found to be employed. This makes it clear that the converse theorem was well established amongst these early geometers.

The other geometrical facts tacitly assumed are:

(1) A circle is the locus of points at a constant distance from a given point. This is made use of in Baudhāyana's first method for the construction of a square by drawing intersecting circles and in the method for drawing perpendiculars with the help of intersecting circles employed by Āpastamba, Baudhāyana and Kātyāyana.

(2) The perpendicular bisector of a line is the locus of points equidistant from the two extremities of the line. This is implied in the construction of perpendiculars by means of intersecting arcs.

(3) The line joining the vertex to the middle point of the base of an isosceles triangle is perpendicular to the base. The

[1] *M. Sl.* p. 6-7.

method of drawing a perpendicular with a string divided into equal parts is based on this.

(4) The tangent to a circle is perpendicular to the radius at the point of contact. In one of the methods (2.8.1) for constructing a square with the help of bamboo rods employed by Āpastamba, the middle point of the side parallel and opposite to the *pṛṣṭhyā* (which itself is one side) is found first, in the course of which two arcs with radii equal to the side of the square and centres at the extremities of the *pṛṣṭhyā* are to be drawn. The next step is to place the bamboo rod with its centre pivoted at the middle point of the side and to adjust it to make its ends touch the circular arcs. The points of contact are the vertices of the square. Here, the operator must have been fully conscious that the bamboo rod placed tangential to the arc will be at right angles to the radius at the point of contact, which is the adjacent side of the square.

(5) A finite st. line can be divided into any number of equal parts. Instances are too many to require quoting.

(6) The diagonal of a rectangle or square bisects it. The construction for reducing a square to an equivalent rectangle with a given side (*Āp. Śl.* III. 1 and *B. Śl.* I.53) and others presuppose the knowledge of this geometrical fact.

(7) The diagonals of a rectangle bisect one another and they divide the rectangle into 4 equal parts, the vertically opposite ones of which are equal in all respects. The knowledge of this truth is evidenced in connection with the construction of the bricks for covering the fire-place. Here the brick got by cutting a rectangular block along one of its diagonals is called an

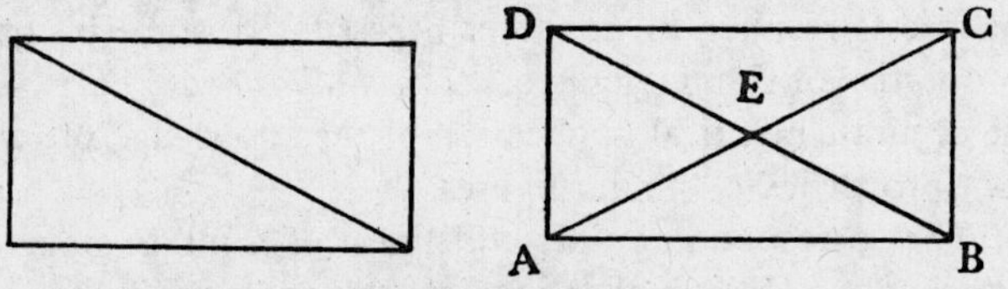

Fig. 28

ardhyā (half-brick), while all the four bricks got by cutting along both the diagonals are termed *pādyā* (quarter) bricks, the acute-angled pair (BEC and AED) being called *śūla-pādyās*[1] (spear-shaped quarter-bricks) and the obtuse angled ones *dīrgha-pādyās* (long quarter-bricks). The authors of the *Śulbaśūtras* were also aware of the fact that these pairs of quarter-bricks when again divided into two equal parts by perpendiculars dropped from the vertex, produced 8 identically equal right triangles.

(8) The diagonals of a rhombus bisect each other at right angles. For converting an *Ubhayataḥ-prauga* (rhombus) into a rectangle, Kātyāyana instructs us to cut the rhombus into two equal isosceles triangles along a diagonal and then proceed as for the *prauga* or isosceles triangle. The procedure for the isosceles triangle implies that the line joining its vertex to the middle point of its base is perpendicular to the base.

(9) The area of an isosceles triangle is equal to half the area of the rectangle with sides equal to the base and altitude of the triangle. This is made use of in the construction of the *praugacit* (the fire-place in the shape of the isosceles triangle). Since altars were not constructed in the shape of scalene and equilateral triangles, we are not in a position to know whether the *Śūtrakāras* realised that the above relation held for all triangles.

(10) The figure formed by joining the middle points of the adjacent sides of a square is itself a square and its area is half the area of the original square. The *Paitṛkī vedī* which is to be a square of area 1 sq. *puruṣa* with its vertices towards the cardinal directions, is constructed by drawing a square of 2 sq. *puruṣas* in the ordinary way and then joining up the middle points of the side.[2]

2.13. *Properties of similar figures*

The enlargement and reduction of the *vedis* and *agnis* practised by the priests resulted in, or rather necessitated, some insight into the properties of similar figures. The *Śulbasūtras* bear witness to the accurate practical application of the knowledge of two important properties of similar figures.

(1) The corresponding sides and lines of similar figures are proportionate. One application of this theorem is particularly interesting. Speaking of the construction of the bricks called

[1] *B. Sl.* III. 167-69 with Thibaut's translation.
[2] *K. Sl.* II. 6.

pakṣeṣṭakas, the bricks to be used in the wing of the *vakra-pakṣaśyenacit* (the altar made like a falcon with curved wings) Āpastamba says:

पक्षकरण्यास्सप्तमं तिर्यङ्मानी । पुरुषचतुर्थं पार्श्वमानी । तस्याक्ष्णयारज्ज्वा करणं प्रजम्भयेत् । पक्षनमन्याः सप्तमेन फलकानि नमयेत् । (XIX. 8)

(The transverse side is $\frac{1}{7}$th of the side of the wing and the lateral side is $\frac{1}{4}$ of a *puruṣa*. Its frame should be expanded diagonal-wise. The planks should be inclined by $\frac{1}{7}$th of the *pakṣanamanī*—the slope or gradient of the wing).

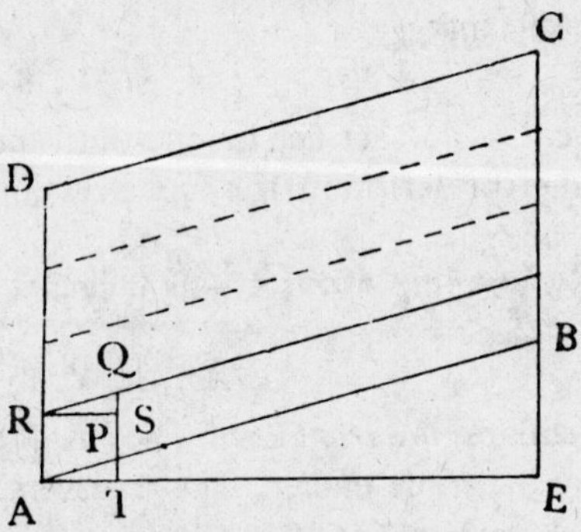

Fig 29

Here what is meant by the *pakṣanamanī* is the vertical height of the tip of the inclined wing from the horizontal. ABCD is half the wing, where CB lies along the vertical. If CB is produced to meet the horizontal through A in E, BE is the *pakṣanamanī*. The bricks which have as their transverse side $\frac{1}{7}$ of AB are to have the same inclination. To effect this the frame for making the bricks is to be given *a namanī* = $\frac{1}{7}$ the *pakṣanamanī*.

i. e. if A P Q R is a brick and Q P is produced to meet the horizontal through A in T we get a triangle A P T similar to the triangle A B E.

$$\therefore \frac{PT}{BE} = \frac{AP}{AB} = \frac{1}{7}$$

$$\text{or } PT = \frac{1}{7} \cdot BE$$

Here in addition to the knowledge of the properties of similar triangles, we can detect rudimentary ideas about trigonometrical ratios of angles.

(**2**) The areas of similar figures are to each other as the square of their sides.

The *Śulbasūtras* record many occasions of decreasing or increasing the size of *vedis* and *agnis* in given proportions. Thus the *Sautrāmaṇī vedi* is to be one-third the *Saumikī* or *Mahāvedi* and the *vedi* for the *Aśvamedha* is to be double the area of the *Mahāvedi*. In the performance of the votive sacrifices, the area of the fire-altar for the first performance is to be $7\frac{1}{2}$ sq. *puruṣas*, and for each subsequent performance, the area is to be increased by one *puruṣa*.[1] But the shape and the proportions of the original *agni* are to be retained religiously. The device adopted to accomplish this is to keep the number of units in the sides unaltered, but to change the size of the linear unit in the ratio of the square roots of the areas. Thus, for the *Sautrāmaṇī vedi* the injunction,for whose area is:

सौमिक्या वेदेर्वितृतीयदेशे यजतेति सौत्रामण्या वेदेर्विज्ञायते ।

(*Āp. Sl.* V. 8)

(It is learnt about the *Sautrāmanī vedi*; sacrifice in one-third the area of the *Saumikī vedi*), the prescription for construction runs:

प्रक्रमस्य तृतीयकरणी प्रक्रमस्थानीया भवति त्रिकरण्या वा, अंष्टिका दशिकेति तिर्यंङ्मान्यौ । द्वादशिका पृष्ठ्या ।

(*Āp. Sl.* V. 8)

(The one-third maker of the (square) *prakrama* will be in the place of the *prakrama*. Or by the triple-maker. The transverse sides will be 8 and 10 and the *pṛṣṭhyā* will be 12.)

Two methods are given here. Either the unit can be taken as $\sqrt{\frac{1}{3}}$ of a *prakrama* and the measurements of the *Saumikī vedi* viz. 24, 30, 36 can be retained, or $\sqrt{3}$ of a *prakrama* may be kept as the unit, when the number of units in the sides will be 8, 10 and 12. In the first case, the area of the *Sautrāmaṇī vedi* is viewed as $\frac{1}{3}$ that of the *Saumikī*, whereas in the second case it is viewed as thrice one-ninth of that area. Besides the knowledge of the properties of similar figures, this equation implies a sturdy grasp of the relation between length and area and the nature of surds.

[1]Eggeling in a footnote on p. 310 (*S.B.E.* Vol. 43) of his translation of the *Śatapatha Brāhmaṇa* says that the intermediate sizes of the fire-altar increase each by 4 sq. *puruṣas* or by one man's length on each side of the body of the altar. But he seems to be labouring under a misconception. Correct interpretation of the passage is difficult unless a person, is acquainted with the contents of the *Śulbasūtras*. And the *Śulbasūtras* are all agreed that the increment should be one sq. *puruṣa*.

The method adopted in *vidhābhyāsa*—increasing the area at every subsequent construction of the *agni*—is based on the same principle. But since the increased size is not a multiple or fraction of the original area, an ingenious modification is introduced. The process as explained by Baudhāyana is:[1]

The excess of the area above the basic $7\frac{1}{2}$ sq. *puruṣas* is made into a square or rectangle, as is convenient. The figure is then divided into 15 equal parts. Two such parts together are turned into a square and combined with a square of one *puruṣa* to yield a new square. The side of this square will be the new linear *puruṣa*, the numerical proportions in the linear measurements being the same as for the basic *agni*. That is if m sq. *puruṣas* is the excess over $7\frac{1}{2}$ sq. *pu*. in any order of the *agni*, $\sqrt{1 + \frac{2m}{15}}$ of a *puruṣa* will be the new unit. Kātyāyana[2] gives more than one way of arriving at $\sqrt{1 + \frac{2m}{15}}$ of a *puruṣa* as the unit of measurement.

One method of enlarging geometrical figures based on the same principle is mentioned even in the *Śatapatha Brāhmaṇa*.

2.14. *Areas*

The *Śulbasūtras* reveal a very clear conception of the relation between lengths and areas. Āpastamba says:

द्वाभ्यां चत्वारि, त्रिभिर्नव।

यावत्प्रमाणा रज्जुस्तावतस्तावतो वर्गान् करोति।

(*Āp. Sl.* III. 6-7)[3]

(With two four, with three nine. As many units as there are in a cord, so many so many squares are produced by it.)

[1] *B. Sl.* II. 12.

[2] *K. Sl.* V. 5, 7, and 10; *S. Br.* X. 2, 3, 7-10.

[3] Compare *K. Sl.* III. 5, 6, 7.

In interpreting *Sūtra* 7, Kapardin omits the repetition of तावत: and takes वर्ग to mean row (*paṅkti*). Following him, perhaps, Thibaut, Bürk and Datta translate the portion तावतस्तावतो वर्गान् as so many rows of squares. But I feel Karavinda's explanation प्रमाणसंख्यागुणितान् प्रमाणसंख्यायुक्तान् वर्गान् is the correct one and consider the later *varga* for square an inheritance from the *Śulbasūtras*.

This knowledge was perhaps the common property of most primitive peoples. But the authors of the *Śulbasūtras* had an equally firm and unerring grasp of the area produced by fractional units of length, which is infinitely more difficult to conceive. For instance, Āpastamba knows a cord of $1\frac{1}{2}$ units produces $2\frac{1}{4}$ units of area and a cord of $2\frac{1}{2}$ units produces an area of $6\frac{1}{4}$ units.[1]

Again $\frac{1}{2}$ a unit produces $\frac{1}{4}$ of a unit and $\frac{1}{3}$ of a unit produces $\frac{1}{9}$th (of a unit of area). A justification or explanation for this is :

अर्धस्य द्विप्रमाणायाः पादपूरणत्वात् ।

(*Āp. Sl.* II. 10)

(Since the half of two units (i.e. one unit) fills up one quarter of the area). Kātyāyana's explanation is :

यावत्प्रमाणा रज्जुर्भवतीति विवृद्धेर्ह्रासो भवति ।

(Since the increase in the area is according to the rule *Yāvat pramāṇā rajju* etc. the decrease should be likewise), i.e. since the multiplication in area produced by a multiplication in the length is that multiple multiplied by itself, the division in area produced by a division in the length should be the latter multiplied by itself. The knowledge of the areas produced by fractional units of length or of the squares of fractions may be the result of observation, but that empirical knowledge was cemented together by inductive logic.

Besides that of the square, the areas of trapezia (only isosceles trapezia are dealt with), isosceles triangles and circles were calculated by the geometers of that age. For calculating the area of a trapezium we are asked to convert the figure into a rectangle[2] from which it is clear that they knew how to calculate the area of a rectangle too.

The *Śulbasūtras* recognise the equalities :

(1) The area of an isosceles trapezium = half the sum of the base and top multiplied by the altitude.

(2) The area of an isosceles triangle = $\frac{1}{2}$ the area of a rectangle whose sides are the base and altitude of the triangle. Also from the construction for the *ubhayataḥ prauga*, twice the

[1] *Āp. Sl.* III 8-9.
[2] *Āp. Sl.* V. 7.

area of an isosceles triangle = ½ the area of a rectangle whose sides are the base and twice the altitude of the triangle.

∴ Area of an isosceles triangle = ½ base x altitude.

(3) The *ardhyā* bricks, right triangles in shape, were halves of rectangles cut diagonally.

Hence the area of a right triangle also

= ½ base x altitude.

We do not know whether the authors of the *Śulbasūtras* had any means of finding the area of a scalene triangle.

(4) The area of a rhombus = ½ the product of the diagonals. This formula is the basis of the construction for transforming an *ubhayataḥ prauga* (rhombus) into a rectangle or a square.[1] For, the method is to cut the rhombus into two isosceles triangles first, then to convert the triangles into rectangles with sides equal to the altitude and half the base, i.e. equal to half the diagonals respectively.

Hence half the area of the rhombus = the product of half the diagonals

∴ Whole area = half the product of the diagonals.

2.15. *Rational right triangles*

With the abundant liberality of an expert, the authors of the *Śulbasūtras* give alternative sets of numbers for forming the right triangle. Thus Āpastamba gives his disciples a choice from five sets[2]—36, 15, 39; 3, 4, 5 multiplied by 4 and 5; 5, 12, 13 and these multiplied by 3; 15, 8, 17; 12, 35, 37.—for constructing the *Mahāvedi*. Baudhāyana gives us a list of right triangles[3] giving the sides about the right angle only. The list comprises :

3,4
12,5
15,8
7,24
12,35
15,36

[1]See 2.10.9 above

[2]*ĀP Sl.* V 2-5.

[3]*B. Sl.* I. 49.

The *Mānavaśulbasūtra* employs some right triangles with sides expressible in fractions for constructing a right angle e.g. $2\frac{1}{2}$, 6, $6\frac{1}{2}$; $7\frac{1}{2}$, 10, $12\frac{1}{2}$. We have already seen how the *Śulbasūtra* method of getting a right triangle on a given side with the other sides as fractions or multiples of the given side leads to one general solution of the rational right-angled triangle viz.

$$2n^2+2n,\ 2n+1 \text{ and } 2n^2+2n+1$$

This gives right triangles whose hypotenuses exceed the longer perpendicular side by one. But the enumerated right triangles do not all conform to this pattern. Kātyāyana's prescription for combining any number of squares of equal size leads to the general relation

$$\left\{\frac{(n-1)}{2}\right\}^2 a^2+n\,a^2=\left\{\frac{(n+1)}{2}\right\}^2 a^2$$

Dr. Datta shows how this relation can yield general formulae satisfying all the rational right triangles occurring in the *Śulbasūtras*.[1] For, this, when $a=1$, gives the relation

$\frac{n-1}{2}, \sqrt{n}, \frac{n+1}{2}$ for the sides of the right triangle.

Putting $n=m^2$, to get rational values, we have

$\frac{m^2-1}{2}, m \ \& \ \frac{m^2+1}{2}$. Here also the difference between the hypotenuse and the longer side is one. If we put $a=2$ in the above relation $(n-1)^2+4n=(n+1)^2$, we get

$$n-1, \sqrt{4n} \ \& \ n+1$$

or (putting $n=m^2$ as before).

$$m^2-1,\ 2m \ \& \ m^2+1 \text{ as the sides.}$$

Here the difference is 2. All the rational right triangles occurring in the *Śulbasūtras* can be brought under these formulae.

Bürk and Thibaut think it improbable that the early mathematicians had any such general formulae as tools. Bürk suggests[2] that all these rational triangles were discovered empirically in the process of constructing larger and larger squares of bricks by the addition of gnomons. The occasions when the gnomons themselves

[1] *Science of the Sulba*, p. 178.
[2] *Z.D.M.G.*, 1901. pp. 565-571.

were capable of being arranged as squares must have been noted, which led to the discovery of various sets of numbers for the right triangle. The explanation is plausible, and this method and other empirical observations might have led to the discovery of a few rational right triangles; but considering that the authors of the *Śulbasūtras* had a genius for generalising (witness Kātyāyana's general rule for combining squares) it is equally possible that they had hit upon some general formulae for deriving the rational right triangle. We also know for certain that they possessed the knowledge that 'k a', k b, k c gives solutions of the right angled triangle, if a, b, c is one solution. For one method for making the corners of the *Saumikī vedi* given by Āpastamba is:

त्रिकचतुष्कयोः पञ्चिकाक्ष्णयारज्जुः, ताभिस्त्रिरभ्यस्ताभिरंसौ चतुरभ्यस्ताभिः श्रोणी ।

(V.3.)

(5 is the diagonal cord for 3 and 4. With these combined with themselves thrice, the top corners are to be marked and with these combined with themselves four times, the bottom corners). i.e. the two right triangles for marking the corners are 4.3, 4.4, 4.5 and 5.3 5.4, 5.5. Similarly in the next *sūtra*, the sides of the right triangle 5, 12, 13 are multiplied by 3 to get another right triangle.

Thus the *Śulbasūtras* are familiar with more than one general formula for finding the sides of rational right triangle, though we are not in a position to know for certain whether they had found the complete general solution.

2.16. *Early Geometrical terminology*

The words *trirasri, caturaśri* and *daśabhuji* which have a geometrical flavour about them, occur in the *Ṛg Veda* and the word *tribhuja* in the *Atharvaveda.* But their exact significance is not known. The *Śulbasūtras* employ a number of technical words.

(1) *Caturaśra* or *Samacaturaśra* stands for square, whereas *dīrghacaturaśra* means a rectangle.[1] These words with the synonym *āyata caturaśra* for a rectangle have been in use ever since.

(2) The word for an isosceles trapezium in the *Śulbasūtras* is *purastād aṁhiyasi* (shorter in front) or *ekato'ṇiṃat* (small to-

[1]The *Nāṭyaśāstra* has a peculiar word vikṛṣṭa (II. 8) for a rectangle, but this word is not found anywhere else in this sense.

wards one end). The Jainas use the word *vetrāsana* (cane seat) or a *vetrāsanasadṛśakṣetra* (figure resembling a cane seat). The parallelogram has seldom received attention.

The rhombus is termed an *ubhayataḥ prauga*, a double isosceles triangle.

(3) The square when viewed as a figure bounded by four equal sides at right angles is termed *samacaturaśra* or simply *caturaśra*. But the unit squares into which any figure had to be divided to compute the area are termed *varga*. This distinction persists on into later days, when also a square figure is a *caturaśra*, but the exponential two i.e. the algebraical square, is *varga*. Area also is *vargaphala*. *Vargakṣetra* is the geometrical representation of a square number just as a rectangle, used to represent the product of two quantities, is termed a *ghātakṣetra*.

(4) The side of a square is called *karaṇī*, producer, which in later mathematics is used for a surd. In the *Śulbasūtras*, though the concept was primarily geometrical, it was used to denote the root of any number square or otherwise. Thus *dvikaraṇī* $=\sqrt{2}$, *trikaranī* $=\sqrt{3}$, *catuṣkaraṇī* $=\sqrt{4}$, $=2$ and so on. But in later mathematics, the roots of square numbers come to be called *mūla* or *pada*, while *karaṇī* is restricted to the roots of non-square numbers. Sometimes the word is applied to the non-square number itself. Its counterpart *kṛti* (the produced) signifies the square of any number and is used as a synonym for *varga*. In the *Pañcasiddhāntikā* (4,5) *karaṇi* seems to be used in the sense of a square.

(5) The diagonal of a square or a rectangle was designated *akṣṇayārajju* or simply *akṣṇayā*, whereas the horizontal and lateral sides were the *tiryaṅmānī* and the *pārśvamānī*. The word *akṣṇayā* (as also the word *prauga* for a triangle) has disappeared from use without leaving a trace. The word has a respectable antiquity occuring in the *Ṛg Veda* itself and being used in the *Brāhmaṇas* in the sense of going across transversely, is well suited to be used for the diagonal. Still in later mathematics we find it completely ousted by a rather senseless word *karṇa* (ear) along with all its synonyms like *śruti*, *śravas*. It is very difficult to understand the semantic basis for this use. Can we fall

back upon the *Nirukta* derivation from *kṛt* to cut, when the *karṇa* will be a line dividing a figure i.e. a diagonal. There is a Greek root, *krino* which means to 'separate, put asunder' which bears a close resemblance to *karṇa* and so supports the *Nirukta* etymology. The root 'karn' to pierce, to hear is probably a denominative from *karṇa*. *Karṇa* used with some geometrical connotation occurs in the *Śulbasūtras* themselves. Kātyāyana after dealing with the conversion of a triangle and a rhombus into a rectangle says:

एतेनैव त्रिकर्णसमासो व्याख्यातः, पञ्चकर्णानां च प्रउगेऽपच्छिद्य ।
एककर्णानां द्विकर्णानां च समचतुरश्रेऽपच्छिद्य ।

(*K. Śl. IV*, 9-10)

(By this itself the combination of *trikarṇas* is explained. The combination of *pañcakarṇas* is by cutting up into triangles. *Ekakarṇas* and *dvikarṇas* are to be combined by cutting up into squares). What figures are denoted by these words? The commentators are of no help in the correct appraisal of their meanings. Dr. Datta translates *pañcakarṇa* as pentalaterals, but finds it difficult to retain the parity *karṇa* = lateral side in *ekakarṇa* and *dvikarṇa*. The plain truth is the word *karṇa* as used in this context still eludes our grasp.[1]

(6) *Prauga* which has no geometrical significance is the word used for a triangle, for an isosceles triangle, which alone makes its appearance in the *Śulbasūtras*. Kātyāyana uses the word *tryaśri* for a triangle but once only (*K. Śl.* iii. 7). The word *tryaśra* is of common occurrence in later mathematics, though *tribhuja* is more popular.

(7) A tell-tale word is *iṣu* used by Kātyāyana only (III. 7) to denote the altitude of a triangle. *Iṣu* means an arrow, but makes its appearance with a geometrical sense quite early, especially in Jaina literature, where, as well as, in later mathematics the word invariably means the height of an arc. This usage is quite understandable if we remember that the arc is termed *cāpa*, *dhanus* etc. all denoting a 'bow'. The use of the word to denote the altitude of an (isosceles) triangle perhaps shows that even at that early

[1]The word *catuḥkarṇa* occurs in the *Jambudvīpasamāsa* (p. 5), and the commentator equates the word to *caturaśra*. Does *karṇa* mean angle or corner and is *koṇa* a prakritisation of *karṇa* ?

date Indian geometry was becoming or had become chord-geometry, which character it retained throughout in later mathematics.

(8) *Maṇḍala* or *parimaṇḍala* denotes a circle, while *pariṇāha*, less commonly used, stands for the circumference. The diameter is *viṣkambha*. All these words are of common occurrence in early Jaina geometry too. But the word *pradhi* for the segment of a circle becomes obsolete in later literature and is not used in early Jaina works. *Parimaṇḍala* seems to be used in the sense of an ellipse in early Jaina literature.

(9) *Kṣetra* primarily means a figure, but sometimes also the area of the figure.

2.17. *The Śulbasūtras and later ages*

There is a charge against Indian mathematics that the earlier phase has no connection with its later phases, especially that the *Śulbasūtra* mathematics has nothing to do with later mathematics. The charge was perhaps first framed by G. R. Kaye,[1] in whose eyes any stick is good enough to beat the Indians with. One could have ignored this charge if it had not been repeated by such a responsible and unprejudiced critic of Indian achievements as A. B. Keith.[2]

A close study of the *Śulbasūtras* in relation to the rest of Indian mathematics will reveal the following facts, out of which a single one only can be adduced as evidence that *Śulbasūtra* mathematics stands apart from the rest of Indian mathematics.

(1) The two most important achievements of *Śulbasūtra* geometry are the enunciation of the theorem of the square on the diagonal and the recognition of the properties of similar figures.

The whole of Indian geometry and trigonometry is dominated by the theorem of the square on the diagonal. The study of rational figures which fills pages in later mathematical texts, the sine-table and even much of algebra is based on this theorem. The field of influence of the principle of proportionality is

[1] *The Source of Hindu Mathematics*, *J.R.A.S.* 1910, p. 749 ff.
[2] A.B. Keith, *A History of Sanskrit Literature*, 1948, p. 517.

even wider, holding sway as it does, over the whole of Indian mathematics.[1]

(2) As far as the constructions in the *Śulbasūtras* are concerned their rightful legatee is the science of architecture. The method for fixing the quarters given in such late works on architecture as the *Tantrasamuccaya* of Nārāyaṇa even, is the same, as the one given by Kātyāyana. The *tantra* texts dealing with the construction of *kuṇḍas* (fire pits) and *maṇḍapas* (halls or sheds) too employ the methods of the *Śulbasūtras.* Thus the *Maṇḍapadruma* of Mahādeva, written in the 17th century[2] makes use of the *Śulbasūtra* method of getting a right angle with cords of length 3, 4 and 5, to construct a square and mentions the *Śulbasūtras.*[3] But an important difference is also noticeable. The mystic importance of the correct measurements no longer holds sway over the minds of the worshippers. Insistence on a high degree of exactness (*saukṣmya*) in the measurements of the *kuṇḍas* and *maṇḍapas* is spoken of as useless.[4] Hence the highly accurate methods of construction of the Vedic altar-makers were more or less a superfluity.

(3) The remarkably close approximation for the value of $\sqrt{2}$ given in the *Śulbasūtras* is apparently lost sight of in later works. But this is nothing strange. The *Śulbasūtra* value is a very good approximation, no doubt, but too cumbersome for manipulation, whereas the method of evaluating surds by multiplying by a large square number such as 10000 yielded very satisfactory results.

(4) All the technical words used in the *Śulbasūtras* whose derivatory meaning could contain a mathematical concept are retained in later mathematics. (There is only one exception to this). Instances are words like *samacaturaśra* (square), *dīrgha-caturaśra* (rectangle), *tryasri* (changed into *tryasra*—triangle), *karaṇi, varga, kṣetra* and the like. On the other hand the word *iṣu* used for the altitude of a triangle in the *Kātyāyana Śulba-*

[1]Bhāskara II himself comments on the dominant role of this principle (*Līl.* 239 and the *Vāsanā* thereunder).

[2]The text has been edited with an introduction by Dr. Sree Krishna Sharma and published in the Adyar Library Bulletin. XXII. 1958 pp. 119-57.

[3]*Maṇḍapadruma*, I. 18-21.

[4]Ibid. I, 2-5.

sūtra seems to be an indication of the early link between Jaina and the rest of Indian mathematics.

(5) The one exception to the statement in (4) is the word for the diagonal of a rectangle, the *Śulbasūtra* word *akṣṇayārajju* being never found again, while the usurper *karṇa* does not seem to have an equally good claim. As against the possibility of this problem of the *akṣṇayā* versus *karṇa* showing a break in our geometrical tradition, we have to remember that the word *karṇa* already appears in the *Śulbasūtras* with a geometrical significance. In the *Sūryaprajñapti* the phrase *karṇakalam* is used in opposition to the word *bhedaghāta* to denote the gradual motion of the sun from one orbit to another[1] i.e. in the sense of 'diagonally'.

Thus there is no sufficient ground for thinking that the *Śulbasūtra* mathematics was forgotten by later ages.

[1]S.P, II, 2.

Chapter III

EARLY JAINA GEOMETRY

3.1. To the period of mathematical development represented by the *Śulbasūtras*, probably belongs the mathematical knowledge of the Jainas too. Dating their canonical literature is as difficult as the dating of the Vedic literature. For long Vardhamāna Mahāvīra, of about the same period as the Buddha, was held to be the founder of Jainism. But Mahāvīra is actually the last of the *Tīrthaṃkaras* and some of the earlier *Tīrthaṃkaras* and *Cakravartins* like Ṛṣabha and Bharata are well known in Hindu Purāṇic literature also. Hence it is more likely that this dissident faith, revolting against sacrificial killing, was quite an old rival to the Vedic faith or that it had taken root in India even before the Vedic faith. The mathematical knowledge contained in the Jaina religious writings should therefore have been more or less parallel to that in the Vedic literature. Whereas the practical necessity of building the altars turned the sacrificing Vedic tribes to mathematics, to geometrical constructions, the Jainas were impelled to indulge in mathematical calculations by an abstract love of precision. Still the connection between the two is easily noticed. The trapezium, the shape of the *Mahāvedi* of the Vedic religion, which in the speculations of the *Brāhmaṇas* attained cosmic significance, reappears in Jaina cosmography as the shape of the universe, the shape of the mountains, and the shape of the *varṣas* (the continents). Along with this geometrical figure, the circle and its parts also came to assume great importance, since the earth and the orbits of the heavenly bodies were given a circular shape. And so the geometry or rather mensuration of the circle and the trapezium constitutes the geometry of the Jaina religious literature.

3.2. The ancient works which are classed as *Gaṇita* texts are the *Sūryaprajñapti*, the *Candraprajñapti* and the *Jambudvīpaprajñapti*. Of these the first two cover more or less the same ground — the peculiar configuration and measurements of the paths of the heavenly bodies worked out into the minutest details. The earth stretches flat with the continents and oceans alternating in concentric circles forming a huge Newton's rings pattern, as it were. The sun and the moon move in concentric circles in planes parallel to the earth, the circles having mount Meru as their centre. The diameter of these circles or *maṇḍalas* goes on increasing gradually. Various views about the diameter and circumference of the *maṇḍalas* are enumerated and rejected by the *Sūryaprajñapti*. According to these the ratio of the circumference to the diameter of a circle[1] i.e. π is 3, while the value accepted by the *S. P.* is the nearer approximation $\sqrt{10}$. This value appears regularly in the Jaina works and even Brahmagupta adopts it.

A few other geometrical figures are mentioned:[2]

Samacaturaśra	—the equal four-sided figure
Viṣamacaturaśra	—the unequal four-sided figure
Samacatuṣkoṇa	—the equiangular quadrilateral i.e. rectangle
Viṣamacatuṣkoṇa	—quadrilateral with unequal angles
Samacakravāḷa	—circle
Viṣamacakravāḷa	—ellipse
Cakrārdha cakravāḷa	—semi-circles

It is interesting to find a distinction being made between the *caturaśra* and *catuṣkoṇa.* The accepted meaning of *asra* or *aśra* or *asri* is a corner and that of *caturaśra* a four-cornered figure which is a quadrilateral. But in the light of the passages in which this distinction occurs *asra* is perhaps to be taken to mean side. Dr. B. B. Datta following Weber translated *samacaturaśra* and *viṣamacaturaśra* as even and oblique square; and *samacatuṣkoṇa* and *viṣamacatuṣkoṇa* as even and oblique parallelogram.[3] The occurrence of the word *koṇa* distinguished as equal and unequal, in such an early work as the *Sūryaprajñapti* indicates that

[1]*Sūryaprajñapti* I. 7 and IV.
This very rough value of π is met with in ancient Babylonian and Chinese mathematics also. The Egyptian value of $(8/9)^2.4$ is very near to the Jaina value of $\sqrt{10}$.

[2]*S.P.* I. 8.

[3]B.B. Datta, The Jaina School of Mathematics, *Bull. of Cal. Math. Soc.* Vol. XXI. 1929.

the Indians had some conception of angles and that the word *koṇa* appearing in later works is of native origin and not an Indianisation of the Greek *gonia*. That the *viṣamacakravāla* (unequal circle, ellipse) should be distinguished from the *samacakravāla*, the equal circle is also interesting more especially when two perpendicular diameters distinguished as *āyāma* and *viṣkambha* are mentioned.[1] These, of course, are equal in a circle.

3.3. The *Jambudvīpaprajñapti* speaks about the arc (*dhanuḥpṛṣṭha*) and the chord (*jīvā*).[2] The arc is also often calculated from the chord, from which we have to conclude that the approximate formulas occuring in the other Jaina works were known to the author of the *Jambudvīpaprajñapti*. The authors of the *Sūryaprajñapti* and the *Candraprajñapti* also could not have been unaware of these, since the two works have the same cosmographical details in their background. The *Jyotiṣkaraṇḍaka* which purports to expound the knowledge contained in the *Sūryaprajñapti*[3] gives the following approximate formulas.[4]

$$c = \sqrt{.4\,h\,(d-h)}$$

$$a = \sqrt{6h^2 + c^2}$$

$$h = \sqrt{\frac{a^2 - c^2}{6}}$$

$$c = \sqrt{a^2 - 6h^2}$$

$$\text{circumference of a cricle} = \sqrt{10d^2}\ (= \pi\, d)$$

$$\text{Area of a circle} = \frac{\text{circumference} \times d}{4}$$

[1]*S.P.*I. 8. In this and other passages where the circumference corresponding to large diameters are calculated with $\pi = \sqrt{10}$ the finding of the square root of large numbers is involved. This is not possible unless there was a system of writing numbers with place value and zero.

[2]As in S. II, p. 67 and S. 16, p. 84, *Sri Jambudvīpaprajñaptyupāṅgam*, Jain Pustakoddhara Fund Series No. 54.

[3]*Jyotiṣkaraṇḍuka Prābhṛta* 1 Sūtra 1.

[4]*Sūtras* 180, 181, 182, 183 and 184. The verse 180 rendered into Sanskrit runs अवगाहोनं विष्कम्भम् उद्गाहसंगुणं कुर्यात् । चतुर्गुणितस्य मूलं मण्डलक्षेत्रस्य अवगाहः॥ The same word *avagāha* occuring in the definition and as the thing defined makes the sense vague. I have followed the commentator Malayagiri in taking the second *avagāha* to mean *jīvā* or chord. The same verse with the last line changed into 'सा जीवा सर्वक्षेत्राणाम्' is quoted in Bhāskara I's commentary on the *Āryabhaṭīya*.

(c is the chord, a the arc, h the height of the segment and d the diameter of the circle)

The *Tattvārthādhigamabhāṣya* (c. 150 B.C.)[1] of Umāsvāti gives, besides these, the formulae.[2]

$$h = \frac{1}{2}(d - \sqrt{(d^2 - c^2})$$

$$\text{and } d = \frac{\frac{c^2 + h^2}{4}}{h}$$

These formulae recur in a number of Jaina religious and semi-religious works, e. g. the *Jambudvīpasamāsa* of Umāsvāti (fourth *Āhnika*), the *Laghusāṅghāyani* of Haribhadra Sūri, which contents itself with the expressions for the circumference and area of a circle (Verse 7).

The first of these, which can be derived from the right triangle formed by the half chord, the radius at its end and part of the radius perpendicular to the chord (i.e. from the $\triangle OAD$ in figure 1), or from the $\triangle CAC'$, requires thorough familiarity with the properties of the right triangle and of circles and chords, for its formulation. The next three are rough approximations from the formulae for the circumference of a circle.

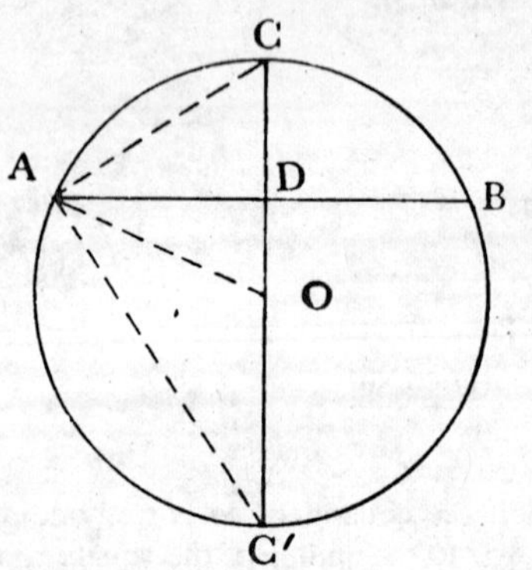

Fig. 1

For, when the segment is a semicircle $c = d$ and $h = \frac{d}{2}$ and so $a = \sqrt{\frac{10d^2}{2}}$ can be written as

$$\sqrt{\frac{4d^2 + 6d^2}{4}}$$

$$= \sqrt{d^2 + 6\left(\frac{d}{2}\right)^2} = \sqrt{c^2 + 6h^2}$$

[1] B.B. Datta, The Jaina School of Mathematics, *Bull. of Cal. Math. Soc.* Vol. XXI. 1929.

[2] *Srīmadumāsvātiviracitam Sabhāṣyatattvārthādhigamasūtram* Ed. by Khubcandraji Siddhantasastri, Raychand Jaina Sastra Mala publication. p. 170.

3.4. There has been much speculation[1] about the origin of the value $\sqrt{10}$ for π, found in all the Jaina works as also in Brahmagupta. The explanation offered by Hunrath seems to be the most plausible, since it is based on the above formula for the chord occuring quite early in the Jaina works. Hunrath's explanation as given by Cantor is :

The arrow of the segment formed by the side of a regular inscribed hexagon will be $h_6 = \frac{1}{2}\left(d - \sqrt{d^2 - \frac{d^2}{4}}\right)$

$$= \frac{d}{4}(2-\sqrt{3})$$

$$= \frac{d}{12} \text{ taking } \sqrt{3} = \frac{5}{3} \text{ approximately.}$$

Now the side of the regular inscribed polygon of 12 sides will be given by $S_{12}^2 = h_6^2 + \frac{1}{4} S_6^2$ (where S_6, S_{12} are the sides of the hexagon and the 12-sided figure respectively.)

$$= \left(\frac{d}{12}\right)^2 + \frac{1}{4}\left(\frac{d}{2}\right)^2$$

$$= \frac{d^2}{144} + \frac{d^2}{16} = \frac{10d^2}{144}$$

$$\text{Hence its perimeter} = 12\sqrt{\frac{10d^2}{144}}$$

$$\text{or } (\text{perimeter})^2 = 10d^2$$

If this is equated to the square of the circumference, π comes out neatly as $\sqrt{10}$.

And this value was quite convenient for the Jaina theologian-mathematician, since the islands and oceans had always diameters measured by powers of ten *yojanas*.

3.5. *Solid figures*

The *Prajñāpanopāṅgam* whose author Śyāmārya or Śyāmācārya died in 92 B.C.[2] says

[1]Quite a few theories have been discussed by M. Cantor in his *Vorlesungen Uber Geschichte der Mathematik*", 4th ed., Vol. I, pp. 647-49.

[2]B.B. Datta, *Jaina School of Maths.*

जे सण्ठाणपरिणया ते पञ्जविहा पण्णप्ता तं जहा परिमण्डलसण्ठाणपरिणया वट्टसण्ठाणपरिणया, तं ससण्ठाणपरिणया, चउरंससण्ठाणपरिणया, आयतसण्ठाणपरिणया।[1]

(*I.S.* 4)

"The arrangements (of the atoms) are said to be of five kinds. That is, elliptically arranged, circularly arranged, triangularly arranged, arranged as a square and arranged like a rectangle".

Commenting on this Malayagiri says that all these are of two kinds, solid (*ghana*) and plane (*pratara*). That is, there are elliptical cylinders, spheres, triangular pyramids, cubes and rectangular parallelopipeds. The minimum number of atoms (or shots) necessary to form these geometrical figures, plane and solid, is also given in each case. To corroborate what he has said, the commentator quotes *karaṇagāthās* in Prākṛt to the same effect. The minimum odd and even numbers of *paramāṇus* with the corresponding configuration are shown in the table on p. 67. The *parimaṇḍala*, it is said, cannot be formed with an odd number of *paramāṇus*. The 20 shots if arranged as directed will form a square with rounded corners rather than an ellipse. In fact it is very doubtful if an ellipse is really meant. The *Bhagavatī-sūtra* mentions all these shapes and numbers.[2] But its commentator Abhayadeva illustrates the *prataraparimaṇḍala* made up of 20 *paramāṇus* by an irregular curvilinear closed figure. Dr. B. B. Datta points out[3] that the *Uttarādhyayanasūtra* and the *Aupapātikasūtras* describe the *īṣat prāgbhāra* as resembling in shape an open umbrella and conjectures the figure referred to as the segment of a sphere. The sūtras say that the thickness of this solid, greatest at its middle, decreases toward the edges at a fixed rate. Hence mensuration of spherical segments was also plausibly familiar to the Jainas.

3.6. *The trapezium and the trapezoidal solid*

The universe, according to the Jainas, is in the form of three trapezoidal solids piled one upon another. As such the

[1]ये संस्थानपरिणताः ते पञ्चविधाः प्रज्ञप्ताः, तद्यथा, परिमण्डलसंस्थानपरिणताः, वृत्तसंस्थानपरिणताः, व्यस्रसंस्थानपरिणताः, चतुरश्रसंस्थानपरिणताः, आयतसंस्थानपरिणताः — is the Sanskrit rendering.

[2]24th Śataka, 3rd Uddeśa, S. 726.

[3]Jaina Mathematics, *Bull. of Cal. Math. Soc.*, Vol. XXI, 1929.

Samsthāna (arrangement)	Pratara		Ghana	
	Minimum odd No of shots	Minimum even No of shots	Minimum odd No of Shots	Minimum even No of Shots
Parimandala (ellipse)		20. (4 in each of the quarters & one in each sub-quarter)		40. 2 layers of 20 each
Vṛtta (circle)	5. 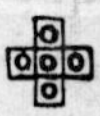	12. 	7. (with 3 in all the 3 dimensions)	32. (a central cube of 8 with 4 projecting on all six sides)
Tryaśra (triangle)	3. 	6. 	35. 5 layers one over another 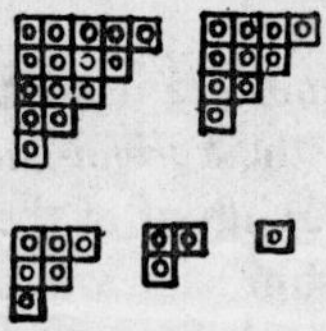	4. 2 layers 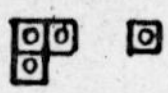
Caturasra (Square)	9. 	4. 	27. (Composed of 3 layers of 9 each.)	8. Composed of 2 layers of 4 each
Āyata (rectangle)	15. 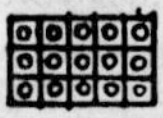	6. 	45. 3 layers of 15 each	12. 2 layers of 6 each

trapezium must have been a familiar geometrical figure in their literature from very early times. In the *Jyotiṣkaraṇḍaka* (10th Prābhṛta V. 190-91) we find a simple rule for calculating the diameter at any height of a mountain shaped like the frustum of a cone. The rate of increase of the diameter is given by

$$\frac{\text{diameter at base—diameter at top}}{\text{height}}$$

Hence the height at which the diameter is required should be multiplied by this ratio and added to the diameter at the top to get the diameter at that height. The geometrical fact that corresponding elements in similar figures are proportionate, known to the authors of the *Śulbasūtras* and the *Brāhmaṇas*, who increased the size of the *vedis* without altering their shape or the proportions of their parts, is apprehended in a more comprehensive and scientific way here. The *Tiloyapaṇṇatti,* which from the quotations from and the references to it in other works must be an early work, has a fuller treatment of trapezia and trapezoidal solids and a fuller application of the properties of similar figures. But since the extant version of the *Tiloyapaṇṇatti* belongs to a much later period, the work is reserved for treatment in a later chapter.

3.7. An important branch of Jaina religio-computational literature, viz., the *karaṇa-gāthās,* are either entirely lost or are yet to be brought out of the obscurity of forgotten private libraries. Stanzas from such works are quoted in almost all Jaina commentaries. What is more, even Bhāskara I, the great exponent of Āryabhaṭa I's mathematics and astronomy, quotes three Prākṛt *gāthās* in his commentary on the *Āryabhaṭīya.* Amongst the mathematicians mentioned by Bhāskara four bear the rather unusual names Maskari, Pūraṇa, Mudgala and Pūtana. The passages in which these names occur are—

एतदेकैकस्य ग्रन्थलक्षया न मस्करि-पूरण-मुद्‌गलप्रभृतिभि: आचार्यैर्निबद्धं कृतं, स कथं अनेनाचार्येण अल्पेन ग्रन्थेन शक्यते[1]

(This has not been composed as a treatise by the *ācāryas* like Maskari, Pūraṇa and Mudgala with even one lakh verses for each (topic). How can this *ācārya* manage it within a short treatise ?)

[1]R. 14850 Govt. Or. Mss. Lib. Madras, p. 2.
[2]Ibid, p. 74.

गणितविदो मस्करीपूरण-पूतनादयः सर्वेषां क्षेत्राणां फलम् आयतचतुरश्रक्षेत्रे प्रत्याययन्ति[1]।

(Mathematicians like Maskari, Pūraṇa and Pūtana show the rationale of the areas of all figures in rectangular figures.) One of these names, that of Pūraṇa,[2] occurs in the *Vyākhyāprajñapti* (3rd *Śataka, Uddeśa* 2). Does this circumstance taken along with the fact that Bhāskara quotes from Jaina works, warrant the assumption that these mathematicians were Jainas ? Be that as it may, it is a pity that mathematicians almost on a par, according to Bhāskara, with Āryabhaṭa and responsible for such voluminous outputs in the field of mathematics should remain mere names for us.

[1]Ibid p. 74.

[2]One *Purāṇa* is mentioned as one of the sages who surrounded Bhiṣma. (*Mahābhārata*, Śāntiparvan, Rājadharmānuśāsana Ch. 47. 12).

CHAPTER IV

THE TRAPEZIUM

4.1. It has been already remarked that the trapezium, more especially the isosceles trapezium, had a place of honour both in the Vedic religion and in the Jaina faith. Though with the ascendency of astronomical mathematics its importance declined a little, it continued to engage the attention of Indian mathematicians right down the ages. The *Śulbasūtras* give geometrical methods for constructing the trapezium and for reducing a square or a rectangle into an isosceles trapezium of the same area. Their authors knew how to compute the exact area of the isosceles trapezium.

4.2. The earliest Jaina work to deal with the trapezium is the *Jyotiṣkaraṇḍaka*, alleged to have been codified at the Valabhi council of the 4th or 6th century. Here the reference is not directly to the trapezium but to the calculation of the diameter at any height of a mountain shaped like the frustum of a cone. Since the vertical section of such a shape will be a trapezium, this amounts to the calculation of the base or top of any section of a trapezuim. The diameter at any height h_1 is given as $d_t + \frac{d_b - d_t}{h} \times h_l$ where d_b and d_t are the diameters at the base and the top and h is the total height of the mountain. Evidently the formula is based on the proportionality of the sides and other elements of similar figures.

4.3. The only verse in the *Āryabhaṭīya* dealing with the trapezium is :

आयामगुणे पार्श्वे तद्योगहृते स्वपातरेखे ते । विस्तरयोगार्धगुणे ज्ञेयं क्षेत्रफलमायामे ।

(*Gaṇitapāda* 8)

(The parallel sides multiplied by the height and divided by their sum give their respective *pātarekhās*. When the height is multiplied by half the sum of the widths, the product should be known as the area in an *āyāma*). This elliptical verse which does not even make it clear that the subject of discussion is the trapezium, is explained by Parameśvara and Nīlakaṇṭha and the translation follows their interpretation. *Pārśve* has to be taken to mean the parallel sides of the trapezium, though such a meaning is unusual for it. The *pātarekhās* are the perpendiculars on the parallel sides from the intersection of the diagonals i.e. EF and EG in the figure. Then $EG = \frac{DC}{AB+DC} \times GF$ and $EF = \frac{AB}{AB+DC} \times GF$ and the area of the figure $= GF \times \frac{AB+DC}{2}$. Of course, this is the usual formula for the area of a trapezium.

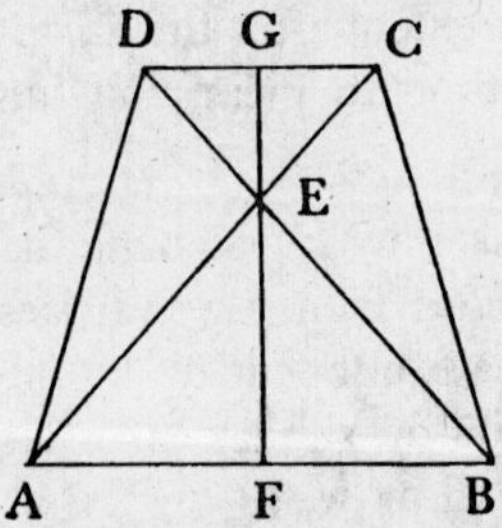

Fig. 1

4.4. Brahmagupta does not give the expression for the area of a trapezium, but notices some other properties.

अविषमचतुरस्रभुजप्रतिभुजवधयोर्युतेः पदं कर्णः ।
कर्णकृतिर्भूमुखयुतिदलवर्गोना पदं लम्बः ॥ (*Br. Sp. Si.* XII. 23)

(In quadrilaterals other than *Viṣama* the square root of the sum of the products of the opposite sides is the diagonal. The square of the diagonal less the square of half the sum of the base and the face is the altitude.)

According to Pṛthūdakasvamin, Brahmagupta's commentator, *aviṣama* includes the square, the rectangle and the isosceles trapezium (*dvisama-samāna-lamba caturbhuja*). The *sūtra* is worded to be applicable to the most general figure of the three, viz., the isosceles trapezium. Let a, b, c, d be the sides of the isosceles trapezium ABCD. Let DE be drawn perpendicular to AB and let BD be joined. Then $BD^2 = DE^2 + BE^2$

$$= AD^2 - AE^2 + BE^2$$

$$= d^2 - \left\{\frac{(a-c)}{2}\right\}^2 + \left\{\frac{(a+c)}{2}\right\}^2$$

$$= d^2 + ac$$

$$= bd + ac \text{ (since } b = d)$$

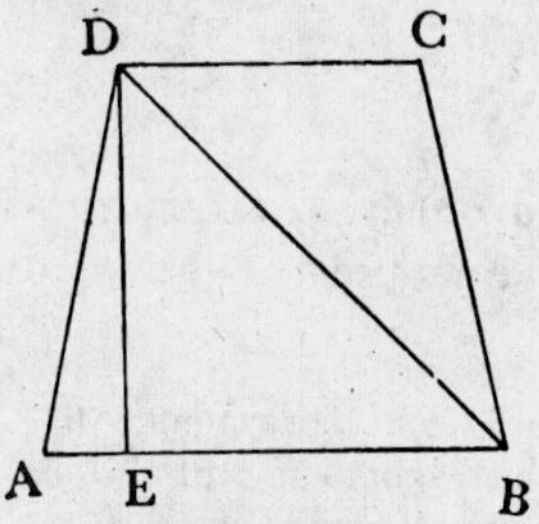

Fig. 2

(In rectangles and squares, since both the pairs of opposite sides are equal,[1] $BD^2 = a^2 + b^2$.)

The expression for the altitude follows directly from the above. Both the formulae are applicable to all *avişama* quadrilaterals except the parallelogram and the rhombus, which incidentally, are two figures which have failed to engage the attention of Indian mathematicians to any considerable extent after the *Śulba-sūtras*. Perhaps *avişama* has to be understood to mean "having the two diagonals not unequal" also.

In the next verse Brahmagupta shows us how to calculate the parts of the diagonals cut off by the altitudes from the vertices of the trapezium and those of the middle altitude cut off by the diagonals. The calculation is based on pairs of similar triangles and is applicable to the general quadrilateral as well.

The expression for the circum-radius of an isosceles trapezium is given in

अविषमपार्श्वभुजगुणः कर्णो द्विगुणावलम्बकविभक्तः हृदयं . . . .

(XII. 26)

(The diagonal of an isosceles trapezium (non-scalene) multiplied by the lateral side and divided by twice the altitude is its circum-radius). For, from the triangle ADB its circum-radius $= \dfrac{BD \times AD}{2DE}$ and this is the same as the circum-radius of the trapezium.

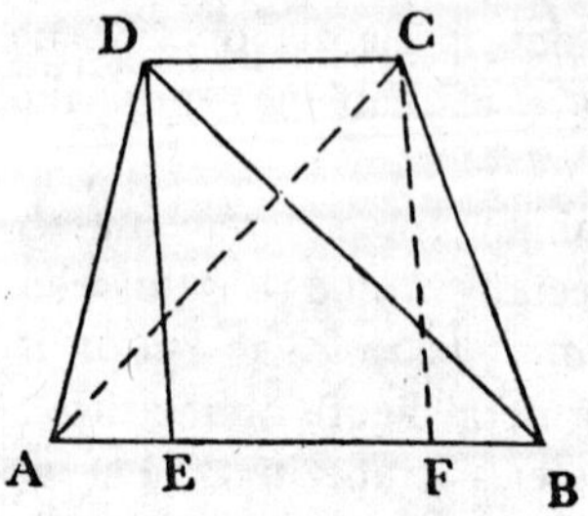

Fig. 3

4.5. If Brahmagupta's use of the word *avişama* for quadrilaterals with equal diagonals is vague, Śrīdhara's use of the word

[1]Pt. Sudhakara Dvivedi in his edition of the *Br. Sp. Si.* says under this verse—"The method given by the *ācārya* gives the diagonal and the altitude in squares and rectangles. In other cases only approximate values of the diameter and altitude are obtained." This is nct justified.

caturasra is still less precise. The first verse in his section on geometry is :

समचतुरश्रायतयोर्भुजकोटिहतिः प्रजायते गणितम्। चतुरश्रेष्वन्येषु च लम्बगुणं कुमुखयोगार्धम् ॥

(*T.S.* 42)

(In a square and a rectangle the product of the base and the upright becomes the area. In other quadrilaterals it is half the sum of the base and the face multiplied by the altitudes). Though the unqualified word *caturasra* is not restricted to the trapezium in its sense, all the three examples given by Śrīdhara under this sūtra are trapezia. The apparent confusion is perhaps the result of a lax use of language. Later on, when he gives a new method for calculating the altitude, he uses the word *ṛju-vadana-caturbāhu* for the trapezium. (The word actually means a quadrilateral with straight or parallel face).

This method is :

ऋजुवदनचतुर्बाहौ मुखोनभूमिर्धरा भवेत्त्र्यश्रे।
तद्बाहू पार्श्वभुजौ प्रकल्पयेन्मध्यलम्बाय ॥
त्र्यस्रस्य फलं द्विगुणं भूभक्तं मध्यलम्बको भवति।
कोटिः स एव बाहुस्त्ववधा कर्णस्तु पार्श्वभुजा ॥

(In a trapezium the base diminished by the face will become the base of a triangle. Its sides will be the flanks of the trapezium. Such a triangle should be constructed for getting the middle altitude. Twice the area of this triangle divided by the base will be the middle altitude. Then that will be the upright side, the segment of the base will be the horizontal side and the flank side will be the hypotenuse (in the right-angled triangle).

The method is useful for a non-isosceles trapezium in which case only the triangle formed according to Śrīdhara's direction will be a scalene triangle. The three sides being known, the area can be calculated with the help of the formula

$$A=\sqrt{s(s-a)(s-b)(s-c)}$$

and thence the altitude and the segments of the base step by step. This method of finding the altitude is not given by other mathematicians.

Śrīdhara's more elaborate work, the *Pāṭīgaṇita* might have contained more geometrical knowledge. Its part, edited by Dr. K.S. Shukla contains an interesting section on *Śreḍhīkṣetras*, diagrammatical representation of series. As in the *Gaṇita Kaumudī* of Nārāyaṇa Paṇḍita arithmetic progressions are represented by isosceles trapezia, their areas being equal to the sums of the series. (*Pāṭīgaṇita* 79-85)

4.6. Mahāvīra lists five kinds of quadrilaterals.

(1) *Sama* with all sides equal — the square and the rhombus.

(2) *Dvidvisama,* with pairs of opposite sides equal — the rectangle and the parallelogram, though the latter does not get any notice in Mahāvīra's work.

(3) *dvisama,* with two sides equal — the isosceles trapezium.

(4) *Trisama,* with three sides equal — the trapezium with three sides equal and

(5) *Viṣama,* with unequal sides, which most frequently, denotes the cyclic quadrilateral. Even the trapezium with unequal sides does not seem to be included under *viṣama*[1] (probably because it is not cyclic). For giving the usual expression for the area of a trapezium Mahāvīra says

भुजयुत्यर्धचतुष्कात् भुजहीनात् घातितात् पदं सूक्ष्मम् ।
अथवा मुखतलयुति दलमवलम्बकगुणं न विषमचतुरश्रे ।

(*G.S.S.* VII. 50)

(The square root from four sets of half the sum of the sides respectively diminished by the sides and multiplied together is the exact area. Or, half the sum of the base and the face multiplied by the altitude, but not in a *viṣama* quadrilateral).[2] From this it is also clear that Mahāvīra knew that the formula

$$A = \sqrt{(s-a)(s-b)(s-c)(s-d)}$$

was applicable to the isosceles trapezium too, though at the same time we have to suspect that he did not realise that the second formula applied to all trapezia, not merely to the isosceles one.

Mahāvīra has a rule for calculating the base and altitude of the sections with proportionate areas into which a given isosceles trapezium is divided by lines running parallel to its parallel sides. The computation is based on the proportionality of the sides of similar triangles.

[1]Compare Brahmagupta's use of *aviṣama* in the sense of a quadrilateral with equal diagonals. The trapezium with unequal sides has apparently no place.

[2]Though Brahmagupta also might have known this he does not give any sure indication that he did so.

If A B C D is a trapezium divided into parts with area in the ratio $\frac{m}{n}$ by the line E P parallel to A B and if D G, the altitude from D is drawn cutting E P in F and meeting A B in G, from the similar triangles D E F and D A G, we have

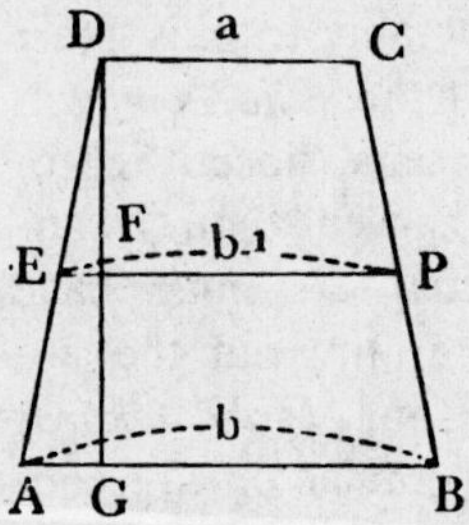

Fig. 4

$$\frac{EF}{AG}=\frac{DF}{DG}$$

i. e. $\frac{b_1-a}{b-a}=\frac{DF}{DG}$, where a and b are the face and the base and b_1 the intermediate base.

$$\text{or } \frac{b_1^2-a^2}{b^2-a^2}=\frac{DF(b_1+a)}{DG\ (b+a)}=\frac{m}{m+n}$$

$$\text{or } b_1^2-a^2=\frac{m}{m+n}(b^2-a^2)$$

$$\text{or } b_1^2=\frac{m}{m+n}(b^2-a^2)+a^2$$

Then the altitude is easily calculated from the area. This solution is given in

खण्डयुति-भक्ततलमुखकृत्यन्तरगुणित-खण्डमुखवर्गयुतम् ।
मूलमधस्तलमुखयुतदलहृतलब्धं च लम्बकः क्रमशः ॥

(*G.S.S.* VII. 175½)

(The part multiplied by the difference of the square of the base and the face and divided by the sum of the parts is combined with the square of the face. The square root of this (is the base). (The area) divided by half the sum of the bottom side and the face is the perpendicular).

Mahāvīra restricts his method to the case where the approximate area of the trapezium is known though it works equally well with the exact area.

4.7. Most of the known Jaina semi-religious works of the post-Christian period, which devote considerable attention to mathematics, belong to a period not far removed from Mahāvīra's date. Vīrasena, author of the *Dhavaḷā Tīkā* on the *Ṣaṭkhaṇḍāgama*, lived in the 8th century, being a contemporary of the Rāṣṭrakūṭa king Jagatuṅgadeva,[1] predecessor of Nṛpatuñga (825 A.D.). The text itself written by Bhūtabali in the first century A.D. contains some mathematical material, though not much of geometry; but the commentator adds considerably to this store of mathematical lore. The extant version of the *Tiloyapaṇṇatti* (*Trilokaprajñapti*) seems to be later than Vīrasena, since it reproduces passages from the *Dhavalā*,[2] but it is earlier than Nemicandra of the 10th century A.D. The original text must have been quite old; but we have no means of separating out the two strata of mathematical accretions. The first four *Mahādhikāras* of this work have a stock of mathematical formulae chiefly connected with the circle, the trapezium and the cylinder ingeo metry, and series in algebra. The *Jambudvīpaprajñapti-saṃgraha* of Padmanandin who lived in Bara in Rajasthan, probably about the close of the 10th century,[3] is similarly based on a very ancient work, the *Jambudvīpaprajñapti*.[4] Nemicandra's (10th century) *Trilokasāra* and *Gommaṭasāra* too serve as store-houses of ancient Jaina mathematical knowledge. Here the *Tiloyapaṇṇatti* is taken as representative of these texts so far as their geometrical knowledge is concerned.

Besides the usual formula for the area of a trapezium (*vetrāsana-sannibha-kṣetra*), the *Tiloyapaṇṇatti* calculates correctly the height of a trapezium at different distances from its lower corners (I. 180). The principle involved is the proportionality of the elements of similar triangles.

[1]Hiralal Jain, आठवी शताब्दी से पूर्ववर्ति गणितशास्त्रसंवन्धी संस्कृत ग्रंथों की खोज- *Jain Antiquary*, Vol. VII, p. 106 ff. But A.N. Singh in his article on the "Mathematics of the *Dhavalā*" published as an introduction to the fourth volume of the *Ṣaṭkhaṇḍāgama* ed. by Hiralal Jain and published by the Jain Sahityoddharaka Fund, Amraoti, places Vīrasena in the 9th century, but thinks the mathematics in the *Dhavalā* is at least four centuries older.

[2]Vide Phul Chand Siddhanta Sastri's "वर्तमान तिलोयपण्णत्ति और उसके रचनाकाल इत्यादि का विचार" published in the *Jain Antiquary*, Vol. X., No. 1.

[3]*Jambudvīpaprajñaptyupāṅgam* ed. by A.N. Upadhye and Hiralal Jain, Introduction, p. 14.

[4]The *Jambudvīpaprajñaptyupāṅgam* published from Bombay does not contain much mathematical material.

Using the same principle the *Tiloyapaṇṇatti* calculates the rate of increase or decrease of the top or base of a trapezium.

भूमिश्र मुहं विसोहिय उच्छेदहिदं मुहाउ भूमिदो ।
सव्वेसु खेत्तेसु पत्तेक्कं वड्ढिहाणीओ ॥

(*T.P.* I 176)

(भूम्या मुखं विशोध्य उत्सेधहृतं मुखाद् भूमीतो ।
सर्वेषु क्षेत्रेषु प्रत्येकं वृद्धिहानी ॥)

(Subtracting the top from the base, the difference is divided by the height. This is the increase and decrease from the top and the base in all (trapezoidal) figures).

The theorem of the square of the hypotenuse is used to calculate the slanting sides of a trapezium from the known base, face and altitude.[1]

4.8. Āryabhaṭa II who was somewhat baffled by the unprecise use of language on the part of his predecessors and was bold enough to voice a protest, criticises Śrīdhara's method of calculating the altitude of a trapezium from the triangle formed with the difference between the parallel sides as base and the flanks of the trapezium as sides, in the following words.

विमुखां धात्री धात्रीं प्रकल्प्य लम्बं करोत्यसौ लम्बः ।
सार्वत्रिकोऽपि न च भूर्नियता तस्मान्मतं तन्न ॥

(*Ma. Si.* XIV. 80)

(Making the base diminished by the face the base, one calculates the altitude. This altitude is not (or, is still ?) the same everywhere. Nor is the base fixed. Hence it is not acceptable to us.)

This criticism itself baffles us by the difficulty of construing the verse. Āryabhaṭa seems to have tried the method for quadrilaterals other than the trapezium and found it unworkable. The school to which Āryabhaṭa II and Bhāskara II belonged seem to have suffered from a lack of continuity of geometrical knowledge, though, to their credit, it must be said that their approach was very rational. Hence perhaps Āryabhaṭa's distinction of not forgetting the parallelogram (and the rhombus) amongst quadrilaterals. He has a rule for calculating the second diagonal of these figures, when one is given.

[1] *Jambudvipaprajñapti* III. 37, *T.P.* II. 179.

समचतुरस्रेऽर्धसमे वाभीष्टश्रवणवर्गोनात् ।
सर्वभुजवर्गयोगमूलं कर्णो द्वितीयः स्यात् ॥

(*Ma. Si.*, XIV 81-82)

(In a rhombus and a parallelogram the second diagonal will be the square root of the sum of the squares of all the sides diminished by the square of the diagonal.)

For if in the parallelogram A B C D the altitudes CE and DF are drawn and the diagonals A C, B D are joined

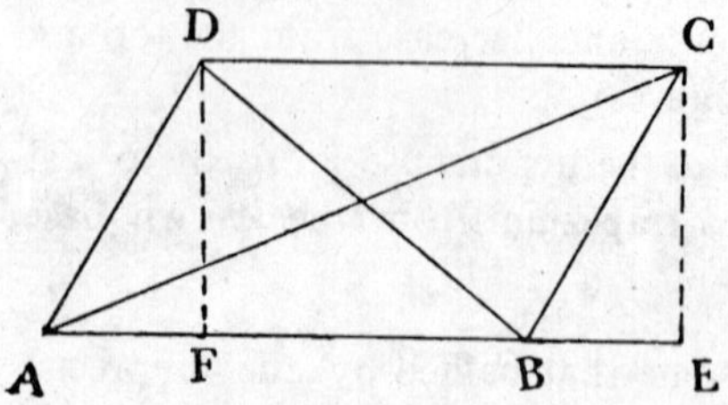

Fig. 5

$$
\begin{aligned}
AC^2 &= AE^2+CE^2 \\
&= (AB+BE)^2+BC^2-BE^2 \\
BD^2 &= BF^2+DF^2 \\
&= (AB-BE)^2+BC^2-BE^2 \\
\therefore AC^2+BD^2 &= (AB+BE)^2+(AB-BE)^2+2\,BC^2-2BE^2 \\
&= 2\,AB^2+2\,BE^2+2BC^2-2BE^2 \\
&= 2AB^2+2\,BC^2
\end{aligned}
$$

$\therefore\ AC^2=2\,AB^2+2\,BC^2-BD^2$ if BD is given

and $BD^2=2\,AB^2+2\,BC^2-AC^2$ if AC is given

The area of a rhombus is given as half the product of the diagonals.

श्रुतिघातः समचतुरस्रे अर्धितः फलं स्यात् ।

(*Ma. Si.* XIV. 82)

(In an equal quadrilateral (square or rhombus) the product of the diagonals when halved, will be the area.)

By implication Āryabhaṭa had the knowledge that the diagonals of a rhombus intersect each other at right angles.

4.9. Śrīpati gives the formula for the area of the trapezium as also the method of calculating the circum-radius of an isosceles trapezium without, however, making it clear that a circum-circle is not possible for the non isosceles trapezium.

चतुर्भुजेष्टश्रुतिपार्श्वबाहुवधस्य चार्द्धं खलु लम्बभक्तम् ।
अतुल्यबाहोः प्रतिबाहुभाग (बाहु ?) वर्गैक्यमूलस्य दलं हि यद्वा ॥

(*Si. Se.* p. 86)

(In a quadrilateral (isosceles trapezium) half the product of a given diagonal and the side adjacent to it divided by the altitude (is the circum-radius). In quadrilaterals with unequal sides it is half the square root of the sum of the squares of the opposite sides). Brahmagupta's expression for the diagonals of an *aviṣama* quadrilateral is also given, बाहुप्रतिबाहुवधैक्यमूलम् (the square root of the sum of the products of the opposite sides). That is, if a, b, c, d are the sides of the *aviṣama*,

$$\text{its diagonal} = \sqrt{ac+bd}$$

Bhāskara II restricts Āryabhaṭa's (II) expression for the diagonal of a parallelogram in terms of the other diagonal and the sides to the quadrilateral with all the sides equal i.e. to the square and the rhombus.

इष्टा श्रुतिस्तुल्यचतुर्भुजस्य कल्प्याथ तद्वर्गविवर्जिता या ।
चतुर्गुणा बाहुकृतिस्तदीयं मूलं द्वितीयश्रवणप्रमाणम् ॥
(*Līl.* 172)

(One diagonal of a quadrilateral with equal sides should be chosen as known. The square root of four times the square of the side decreased by the square of that (diagonal) is then the measure of the second diagonal.)

One definite improvement in the terminology used by Bhāskara is that he has a definitive name for the trapezium viz. *samalamba-caturbhuja* a quadrilateral with constant altitude. The formula for its area is the usual one. For the demonstration of its validity in the case of a non-isosceles trapezium, Bhāskara recommends division into three parts along the altitudes from the two corners with the least perpendicular distance between them, when the middle segment will be a rectangle and the other two segments will be unequal right triangles, all the three having the same upright side. The areas are calculated separately and added up to show that the total tallies with the area as calculated with the help of the formula, $A = \frac{1}{2}$ the sum of the base and face multiplied by the altitude. The demonstration is purely empirical.

4.10. Unlike Āryabhaṭa II, Bhāskara sees the usefulness of the device of forming a triangle with base equal to the difference of the parallel sides and sides equal to the flanks of a trapezium for finding its altitude and *ābādhās*.

समानलम्बस्य चतुर्भुजस्य मुखोनभूमिं परिकल्प्य भूमिम् ।
भुजौ भुजौ त्र्यस्रवदेव साध्ये तस्याबाधे लम्बमितिस्ततश्च ॥

(*Līl.* 184)

(Taking the base of the trapezium diminished by its face as the base and its flank sides as the sides, its (of the triangle so formed) *ābādhās* are to be computed as in (other) triangles and then the measure of the altitude from that.)

Though Bhāskara has already given Āryabhaṭa's condition for any four lengths to enclose a space, he investigates the possibility of their forming a trapezium and gives the condition
समानलम्बे लघुदोःकुयोगान्मुखान्यदोःसंयुतिरल्पिका स्यात् ।

(*Līl.* 185)

(In a trapezium the sum of the other flank side and the face is smaller than the sum of the smaller flank side and the base). Gaṇeśa (16th century) attempts a proof for this by logical reasoning.[1] But the reasoning itself is confused.

4.11. Nārāyaṇa Paṇḍita gives the rules for the calculation of the area, the altitude, the diagonal and the *ābādhās* in a trapezium. Mahāvīra's method for dividing a trapezium into sections with areas in a given proportion by lines parallel to the parallel sides occurs (in *G. K. Kṣetravyavahāra* iii), besides a rule for calculating the middle bases (*madhyabhūmi*) when a trapezium is divided into any number of parts by lines parallel to the parallel sides at regular intervals from the base (*G.K.*, *Ks. Vya.* 21). As in Śrīdhara's *Pāṭīgaṇita*, trapezia are used to represent arithmetic progressions. These may have negative bases or faces and then the flanks cross over each other. In the *Gaṇita Kaumdī* such trapezia also occur in connection with the construction of rational trapezia with base and diagonals equal[2] to one another and of rational trapezia with a given area.[3] Nārāyaṇa and Śrīdhara are the only two amongst known authors in whose works such trapezia make their appearance and the question of the source of this new concept is intriguing.

The later Āryabhaṭa School does not pay much attention to the trapezium.

[1]*Līl.*, p. 175.
[2]Notes under verses *G.K.*, *Ks. Vya*, 88-90.
[3]Notes under verses *G.K.*, *Ks. Vya*, 108-09.

CHAPTER V

THE QUADRILATERAL

5.1. Two sections of Indian mathematicians have approached the study of the quadrilateral from two different angles. One section viewed it merely as a figure enclosed by four chords of a circle, whereas the other viewed it as a figure enclosed by any four lines i. e. the general quadrilateral but, strangely enough, excluding the cyclic quadrilateral. The former includes the majority of Indian mathematicians, Brahmagupta, Śrīdhara, Mahāvīra, Śrīpati and the later Āryabhaṭa School; the latter Āryabhṭa II and Bhāskara II. We do not know to which camp Āryabhaṭa I belonged, since his extant works accord the quadrilateral a doubtful passing notice only, but it is likely he belonged to the first camp. Nārāyaṇa Paṇḍita has leanings towards both. He sympathises with Āryabhaṭa II's criticism of Brahmagupta's theorem for the area of a quadrilateral, evidently because he is not aware that Brahmagupta meant it to be of restricted applicability only. But at the same time he deals with the cyclic quadrilateral too and has discovered some of its properties which apparently Brahmagupta did not know, but were common knowledge in the later Āryabhaṭa School.

5.2. Some of the rules of calculation given by Brahmagupta and his followers, though primarily intended for the cyclic quadrilateral, are applicable to the general quadrilateral as well. Such are

(1) विषमचतुरश्रमध्ये विषमत्रिभूजद्वयं प्रकल्प्य पृथक् ।
कर्णद्वयेन पूर्ववदाबाधे लम्बकौ च पृथक् ॥ (*Br. Sp. Si.* XII. 29)

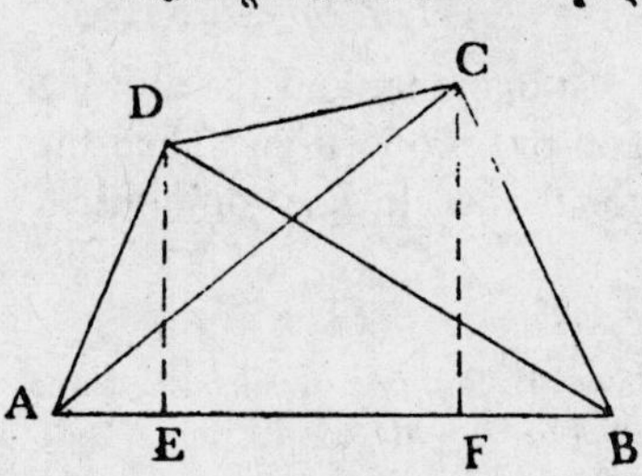

Fig. 1

(Making two scalene triangles in a scalene quadrilateral separately with the two diagonals, one should calculate the altitudes and *ābādhās* as before) i.e. by considering the triangles A B C and ADB (Fig. 1.) C F, B F and D E, A E can be calculated.

2. विषमभुजान्तस्त्रिभुजे प्रकल्प्य कर्णौ भुवौ तदाबाधे ।
पृथगूर्ध्वाधरखण्डे कर्णयुतौ कर्णयोरधरे ॥
त्रिभुजे भुजौ तु भूमिस्तल्लम्बो लम्बकाधरं खण्डम् ।
ऊर्ध्वमवलम्बकखंडं लम्बयोगार्धमधरोनम् ॥

(*Br. Sp. Si.* XII, 30-31)

(In the triangles inside a scalene quadrilateral consider the diagonals as the bases and (calculate) the *ābādhās* which are the upper and lower parts of the (diagonals?) separately. In the triangle beneath the intersection of the diagonals, the lower parts of the diagonals are the sides and the base of the quadrilateral is the base. Its altitude will be the lower segment of the altitude (of the quadrilateral) through the point of intersection of the diagonals and the upper segment will be half the sum of the two (side) altitudes diminished by the lower segment.)

The wording of the first verse is vague and we do not know whether the *ābādhās* themselves are the upper and lower segments of the diagonals (—this is the interpretation accepted by S. Dvivedi, the editor of the *Br. Sp. Si.*), in which case the rule will apply to one particular type of the cyclic quadrilateral only, viz. the one in which the diagonals cut at right angles.

5.3. *Sūcīkṣetras*

In connection with the quadrilateral, most Indian texts speak about a *sūcī* i.e. the triangle formed by producing the flanks till they meet. Brahmagupta shows how to calculate the segments of the altitude of such a triangle and the segments of the diagonals of the quadrilateral produced by their point of intersection—

कर्णावलम्बकयुतौ खण्डे कर्णावलम्बयोरधरे ।
अनुपातेन तदूने ऊर्ध्वे सूच्यां सपाटायाम् ॥

(*Br. Sp. Si.* XII. 32)

(In the *sūcī*,[1] the lower segments of the diagonal and the altitude made by their intersection can be got by proportion. Then the upper segments are (the whole diagonal and altitude) diminished by these (lower segments).

[1]I have left the word *sapāṭāyām* untranslated. Even if *pāṭa* means the same as the *pīṭha* of Bhāskara II and Nārāyaṇa, as S. Dvivedi says it does, I do not see the relevance of the word in the context.

In the figure, E A B is the *sūcī* of the quadrilateral A B C D. E F is the altitude (*sūcīlamba* as it is often called) which intersects the diagonals A C and B D at P and Q respectively. The verse indicates the method for calculating E P, P F and E Q, Q F as also P A, P C and Q B, Q D. If D G and C H the altitudes at D and and C are drawn, we get two sets of similar triangles C A H, P A F and D B G, Q B F from which P F, Q F and P A, Q B can be calculated.

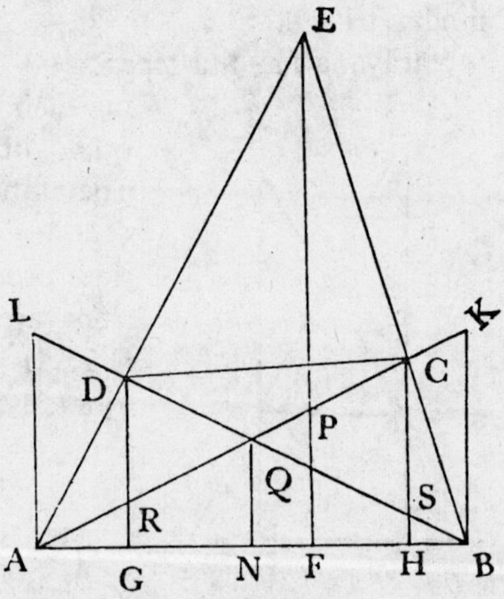

Fig. 2

Bhāskara devotes six verses to a consideration of the *sūcikṣetra* giving (1) expressions for calculating R G, R A and S H, S B (fig. 2) in terms of the *sandhis* A G and H B (the *sandhi* is the projection of the side on the base), the *pīṭhas* G B and AH (the *pīṭhas* are the portions of the base after the *sandhi)* and the end altitudes C H and D G (*Līl.* 193-194 and *Vāsanā* thereon) (2) expressions for calculating the *vamśas*, A L and B K (*vamśas* are the uprights at A and B meeting the diagonals produced at L and K) in terms of the end altitudes, the base and the *pīṭhas* (*Līl.* 195), (3) a direction for calculating the altitude from the point of intersection of the diagonals M, and the segments of the base A N, B N made by it in terms of the *vamśas* and *lambas* (*Līl.* 195) (4) expressions for calculating the *ābādhās* of the *sūcī*, A F and B F; the *sūcīlamba* E F; and the sides of the *sūcī* E A and E B (*Līl.* 196-198).[1]

[1]Here Bhāskara introduces certain new technical terms like *sama* (the *sañdhi* divided by its altitude and multiplied, by the other) for mere facility of expression. This *sama* combined with the opposite *sandhi* is called *hara* (the divisor) but this is because this quantity appears as the divisor in the expressions to follow.

All these expressions are derived from appropriate sets of similar triangles.

Nārāyaṇa Paṇḍita repeats all Bhāskara's calculations swelling their bulk at the same time by another but exactly similar calculation. The altitudes at the ends viz. D G and C H are produced to meet the opposite sides produced at X and Y and expressions are derived for the lengths of G X, H Y and B X, A Y. (*G. K. Ks. Vya.* 58-59).

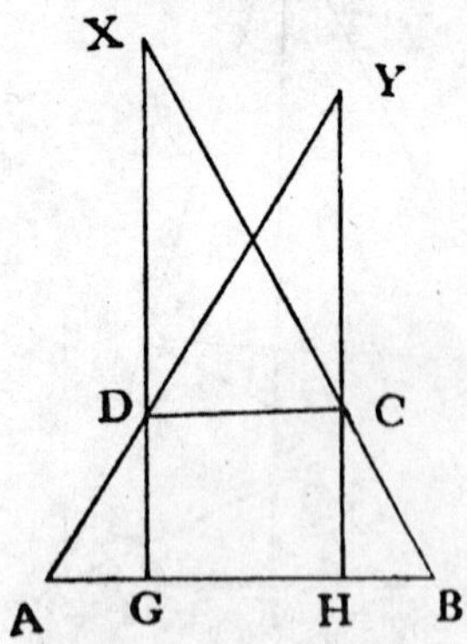

Fig. 3

The concept of *sūcīkṣetras* perhaps first arose in connection with problems of excavation, it being often necessary to take into account the volume of a pit or a tank if it were to taper to a point, before the actual volume of the excavation could be calculated and a pit or a tank generally has a trapezium as cross section. Later on perhaps the field of *sūcīkṣetras* was extended to cover irregular quadrilaterals.

5.4. Noncyclic quadrilaterals

Amongst the known Indian mathematicians Āryabhaṭa II was the first to question the validity of Brahmagupta's expression for the area of a (cyclic) quadrilateral. Perhaps due to a lack of coordination of mathematical knowledge in the different schools, the true significance of the theorem failed to be grasped by Āryabhaṭa II and by Bhāskara II following him closely. Though this was regrettable in itself, it had a salutary effect also, in that it led to a consideration of the general quadrilateral, which had been hitherto left untouched.

5.4.1. Āryabhaṭa II gives a formula for the approximate area of a quadrilateral after giving a general formula for the exact area of a trapezium, triangle, square and rectangle.

त्रिभुजचतुर्भुजफलानयनाय
वदनक्षितियोगदलं लम्बहतं जायते गणितम् ।
त्रिभुजे समचतुरस्रेऽर्धसमे वा कर्णभेदेऽपि ॥
शृंगाटके न नियमाद्विषमचतुर्बाहुके च न प्रायः ।
याम्योत्तरलम्बैक्यार्द्धं क्वा (कु+आ) स्यैक्यार्धताडितं निकटम् ॥

(*Ma. Si.* XIV. 78-79)

(For calculating the area of a triangle and a quadrilateral—Half the sum of the face and base multiplied by the altitude is the area in a triangle, a square, a half equal quadrilateral (one with two sides equal i.e. the rectangle and isosceles trapezium) and also when the diagonals differ (i.e. in a trapezium with unequal diagonals). This never gives the area of a *śṛṅgāṭaka*[1] nor that of most *viṣama* quadrilaterals. Half the sum of the southern and northern altitudes multiplied by half the sum of the base and face is the approximate area (in these).

The last of these rules is the easy method of multiplying the mean altitude by the mean base. For the exact area, the areas of the triangles into which a diagonal cuts a quadrilateral, are to be computed separately and added.

चतुरस्रेऽन्नत्यत्रिभुजद्वयफलयुतिर्गणितम् ।
तत्नत्यकयोः कर्णो भूः स्यादितरे भुजाश्च चत्वारः ॥

(*Ma. Si.* XIV. 68)

(In that (*viṣama*) quadrilateral, the area is the sum of the areas of the two triangles in it. Their base is the diagonal and the other sides are the four sides of the quadrilateral).

5.4.2. For calculating the diagonals, the altitudes at the ends of the diagonals are used, when the calculation reduces to a

[1]What exactly is meant by *śṛṅgāṭaka ?* Monier Williams in his *Sanskrit-English Dictionary* gives one geometrical meaning only for the word and that is triangle. Sometimes it is used in the sense of a tetrahedron. But here it does not seem to denote either. In XIV. 75. Āryabhaṭa speaks of a *śṛṅgāṭakacaturasra* also.

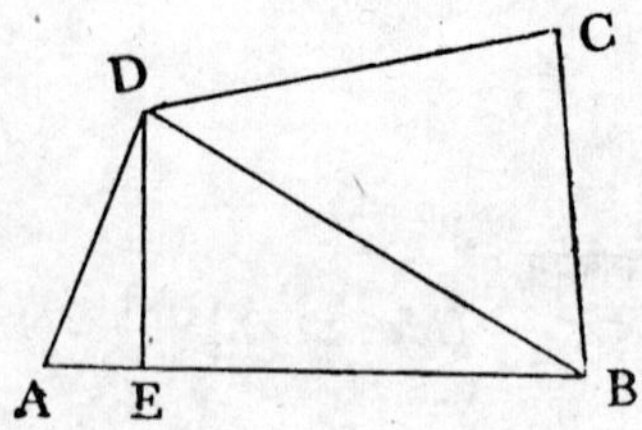

Fig. 4

straight-forward application of the theorem of the square of the diagonal.[1] For, the projection

$$AE = \sqrt{AD^2 - DE^2}$$

and then

$$BD = \sqrt{BE^2 + DE^2}$$
$$= \sqrt{(AB - AE)^2 + DE^2}$$

($BE = AB + AE$, if the altitude falls outside the quadrilateral).

For finding the second diagonal when the first is given

इष्टश्रवणं वसुधां परिकल्प्य त्रिभुजयोरुभयोः ।
लम्बाबाधाः साध्याः स्थाप्य कर्णाग्रमूलयोर्बाधाः ।
क्षेत्रद्वयबाधान्तरवर्गाल्लम्बैक्यवर्गयुतात् ।
मूलं द्वितीयकर्णश्चतुरश्राणां च सर्वेषाम् ॥

(*Ma. Si.* XIV. 86-87)

(Making the given diagonal the base in both the triangles (formed by it in the quadrilateral) the altitudes and *ābādhās* are to be calculated. The *ābādhās* are to be placed in sets touching the bottom and top ends of the diagonal. The square root of the square of the difference between (corresponding) *ābādhās* in the two triangles combined with the square of the sum of the altitudes gives the second diagonal in all quadrilaterals).

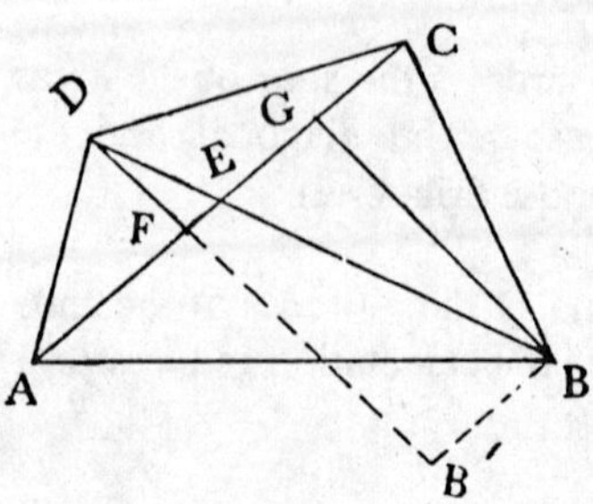

Fig. 5

The necessary lines are shown in the figure (fig. 5). If the altitude DF is produced and B B′ is drawn perpendicular to it from B

$$BB' = GF$$
$$= AG - AF$$

and $FB' = BG$.

∴ From right angled triangle DB′B

$$BD^2 = B'D^2 + BB'^2$$
$$= (DF + BG)^2 + (AG - AF)^2$$

[1] *Ma Si.* XIV. 83-85.

5.4.3. Bhāskara II repeats all these expressions[1] except that for the approximate area. In addition, he investigates the condition that has to be satisfied by the diagonals of a quadrilateral.

कर्णाश्रितस्वल्प-भुजैक्यमुर्वीं प्रकल्प्य तच्छेषभुजौ च बाहू ।
साध्योऽवलम्बोऽथ तथान्यकर्णः स्वोर्व्याः कथंचिच्छ्रवणो न दीर्घः ।
तदन्यलम्बान्न लघुस्तथेदं ज्ञात्वेष्टकर्णः सुधिया प्रकल्प्यः ।।

(*Lil.* 182)

(With the smaller of the sums of the sides about the diagonal as base and the other sides as sides the altitude is to be calculated, similarly the other diagonal. The other diagonal cannot in any case be longer than its base (i.e. the base of the triangle formed as above), nor can the first be shorter than its altitude......)

The process of arriving at these conditions is explained by Bhāskara himself. If the quadrilateral is pressed diagonally, at the extreme position, the two sides at one end of the diagonal coincide with the other diagonal. Hence the two conditions.

All these new rules are prefaced by Āryabhaṭa with a vehement attack on earlier mathematicians.

कर्णज्ञानेन विना चतुरश्रे लम्बकं फलं यद्वा ।
वक्तुं वांछति गणको योऽसौ मूर्खः पिशाचो वा ।।

(*Ma. Si.* XIV. 70)

(The mathematician who wants to find the altitude or area of a quadrilateral without knowing (the length) of a diagonal is a fool or a devil).

Āryabhaṭa had discovered that the quadrilateral, unlike the triangle, is not determined by the sides only. But we do not know whether he was aware that any five elements, not necessarily including a diagonal, will determine a quadrilateral. Anyway, Bhāskara who says

लम्बयोः कर्णयोर्वैकमनिर्दिश्यापरं कथं ।
पृच्छत्यनियतत्वेऽपि नियतं चापि तत्फलं ।
स पृच्छकः पिशाचो वा वक्ता वा नितरां ततः ।।

(*Vāsanā* under *Lil.* 17)

(Without specifying one of the altitudes or diagonals how can any one ask for its determinate area, when it is indeterminate?

[1]*Lil.* 179-81.

The one who asks and the one who replies are fools or devils...), certainly knew this. Again Āryabhaṭa rejects without any compunction Brahmagupta's formula for the area of the cyclic quadrilateral, whereas Bhāskara gives it with the qualification "अस्फुटफलं चतुर्भुजे" (this is the approximate area in quadrilaterals).

Nārāyaṇa Paṇḍita follows Bhāskara closely in his treatment of the quadrilateral, though he does not condemn Brahmagupta's theorem for the diagonals of a cyclic quadrilateral, which he repeats without Bhāskara's apology; but still he does not say specifically that the theorem is true in its sphere, i.e. for the cyclic quadrilateral. He notes that the area calculated with the formula $A = \sqrt{(s-a)(s-b)(s-c)(s-d)}$ is not exact in some quadrilaterals but fails to define these.

5.5. *The cyclic quadrilateral*

In fact Brahmagupta's theorem about the cyclic quadrilateral must have been a hard nut to crack for quite a few earlier Indian mathematicians. Consequently the cyclic quadrilateral has had a chequered history in India.

5.5.1. Brahmagupta's most important contributions to the geometry of the cyclic quadrilateral are the two theorems

स्थूलफलं त्रिचतुर्भुजबाहुप्रतिबाहुयोगदलघातः ।
भुजयोगार्धचतुष्टयभुजोनघातात् पदं सूक्ष्मम् ॥

(*Br. Sp. Si.* XII. 21)

(The gross area of a triangle or a quadrilateral is the product of half the sums of the opposite sides; the exact area is the square root of the product of four sets of half the sum of the sides (respectively) diminished by the sides)

and

कर्णाश्रितभुजघातैक्यमुभयथान्योन्यभाजितं गुणयेत् ।
योगेन भुजप्रतिभुजवधयोः कर्णौ पदे विषमे ॥

(*Br. Sp. Si.* XII. 28)

(The sums of the products of the sides about the diagonal should be divided by each other and multiplied by the sum of the products of the opposite sides. The square roots of the quotients are the diagonals in a *viṣama* quadrilateral).

The first verse embodies the formula for the area of a cyclic quadrilateral

$$A = \sqrt{(s-a)\ (s-b)\ (s-c)\ (s-d)}$$

and the one for the area of a triangle

$$A = \sqrt{s\ (s-a)\ (s-b)\ (s-c)}$$

The one for the triangle is found earlier in the *Metrica* of Heron of Alexandria (between 2nd century B.C. and 3rd century A.D.). Still the case for borrowing by Brahmagupta is very weak. In India the result was derived for the quadrilateral first and then extended to cover the triangle or rather the triangle was stretched into a quadrilateral to bring it within the purview of the formula. Secondly and consequently, Brahmagupta's mode of derivation of the formula is entirely different from the Greek mode, if the derivation and proof given by the Ārya-bhaṭa School gives a clue to his procedure[1] (this proof will be dealt with later).

The second verse quoted above contains the theorem about the diagonals of a cyclic quadrilateral, now generally known as Brahmagupta's theorem, and rediscovered[2] in Europe in 1619 A.D. by W. Snell.

If a, b, c, d are the lengths of the sides of a cyclic quadrilateral and x and y are its diagonals,

the theorem states $x = \sqrt{\dfrac{(ab + cd)\ (ac + bd)}{ad + bc}}$

and $y = \sqrt{\dfrac{(ad + bc)\ (ac + bd)}{ab + cd}}$

thus enabling the diagonals to be calculated in terms of the sides.

[1]In spite of the apparent rivalry between Āryabhaṭa I and Brahmagupta, the latter's theorems and results are preserved and developed in the Ārya-bhaṭa School. One has sometimes a suspicion that these formulae had been current knowledge in Āryabhaṭa's school even before Brahmagupta. Like Bhāskara not understanding and criticising some of Brahmagupta's results but at the same time accepting the bulk of the latter's mathematics, Brahmagupta too, probably drew on Āryabhaṭa I's mathematics largely, adding his criticism wherever he thought it necessary.

[2]D.E. Smith-*History of Mathematics*, Vol. II p. 286.

5.5.2. A third significant contribution by Brahmagupta to the study of the cyclic quadrilateral is his method for getting a rational cyclic quadrilateral, which is to multiply the sides of two rational right triangles by each other's hypotenuse and use these sides as the sides of the quadrilateral. What Brahmagupta obviously does here is to get two pairs of sides, such that the sum of the squares of the sides in one pair is equal to the sum of the squares of the sides in the other. And hence his formula for the circum-radius of a quadrilateral, which is meant to cover such quadrilaterals only.

हृदयं विषमस्य भुजप्रतिभुजकृतियोगमूलार्धम् ॥

(*Br. Sp. Si.* XII. 26)

(The circum-radius of a *viṣama* quadrilateral is half the square root of the sum of the squares of opposite sides).

Since Brahmagupta, like all other Indian *ācāryas* of yore, has left out the logic and the rationales of the results garnered in his work, the interpretation of this line was a challenge for some time. (Nārāyaṇa Paṇḍita of the 14th century criticises the rule saying it does not cover all cyclic quadrilaterals (*G. K.* Part II p. 175). Sudhakara Dvivedi, editor of the *Brahmasphuṭasiddhānta*, commenting on it, says that "according to Brahmagupta the circum-radius could be known only in the case of a cyclic quadrilateral in which the arcs on the opposite sides added up to the semicircumference". But actually the scope of the formula is a bit wider. An anonymous commentator of Nārāyaṇa's *Tantrasāra*[1] gives a clue for the correct appreciation of Brahmagupta's rule. According to him one method of forming Brahmagupta's rational cyclic quadrilateral is to juxtapose two rational right triangles having the same hypotenuse, with their hypotenu-

[1]pp. 265-66 in the transcript of a *grantha* belonging to Desamangalath Nampūtiripāḍ and containing the text of the *Tantrasaṅgraha* with a Malayālam commentary and a Malayālam commentary on the third and fourth chapters of another astronomical text, the *Tantrasāra* composed by Nārāyaṇa of Perumanagrāma. In between the two there is a section which deals with diverse mathematical topics and quotes from the *Līlāvatī* often and from the *Yuktibhāṣā* (p. 268). This explanation of the formation of the rational cyclic quadrilateral occurs in this portion. The transcript used by me belongs to Śri Rāma Varma (Maru) Thampuran, Sivapadam, Marar Road, Trichur.

ses coinciding. Then that hypotenuse will be a diagonal of the quadrilateral as also a diameter of the escribed circle. By interchanging the sides, which are chords, other cyclic quadrilaterals are obtained. But, in all these, either the diagonals will cut at right angles or else one of the diagonals will be a diameter of the escribed circle. Moreover, in these quadrilaterals the sum of the squares of the largest and least sides will be equal to the sum of the squares of the remaining sides.

Brahmgupta's direction for forming the rational quadrilateral is

जात्यद्वयकोटिभुजाः परकर्णगुणाः भुजाश्चतुर्विषमे ।
अधिको भूर्मुखं हीनो बाहुद्वितयं भुजावन्यौ ॥

(*Br. Sp. Si.* XII. 38)

(The *koṭis* and *bhujas* of two *jātyas* multiplied by each other's hypotenuse are the four sides in a *viṣama* quadrilateral. The longest is the base, the least the face and the remaining two sides the flanks).

Hence his formula for the circum-radius is valid if the diagonals cut at right angles. It can be extended to the second case of the diagonal coinciding with the diameter of the circle by taking half the square root of the sum of the squares of the longest and shortest sides.

Speaking about Brahmagupta's geometry Cajori says[1] "Remarkable is Brahmagupta's theorem on cyclic quadrilaterals

$$x^2 = \frac{(ad + bc)(ac + bd)}{ab + cd}$$

and $$y^2 = \frac{(ab + cd)(ac + bd)}{ad + bc}$$ where x, y are the

diagonals and a, b, c, d the lengths of the sides; also the theorem, that if $a^2 + b^2 = c^2$, and $A^2 + B^2 = c^2$, then the quadrilateral (Brahmagupta's quadrilateral) aC, cB, bC, cA is cyclic and has its diagonals at right angles". Actually the significance of the latter theorem is wider, any quadrilateral whose sides are aC, bC, Ac, Bc taken in any order being cyclic. Its diagonals cut at right angles if the greatest and least sides are placed opposite to each other. Moreover, if the diagonals do not cut at right

[1] *A History of Mathematics*, New York, 1919, p. 87.

angles, one of them equal to cC will pass through the centre of the escribed circle. Thus the circum-diameter of such quadrilateral is easily known. The formula for the circum-diameter of the general cyclic quadrilateral is the same as the one for the circum-diameter of the triangle formed by joining a diagonal.

त्रिभुजस्य वधो भुजयोर्द्विगुणितलम्बोद्धृतो हृदयरज्जुः ।
सा द्विगुणा त्रिचतुर्भुजकोणस्पृग्वृत्तविष्कम्भः ॥

(*Br. Sp. Si.* XII. 27)

(The product of the two sides other than the base divided by twice the altitude is the circum-radius of a triangle. Twice that is the diameter of the escribed circle of the triangle and the quadrilateral). Again Brahmagupta fails to specify that the triangle is the triangle formed by joining the diagonal of a cyclic quadrilateral.

Brahmagupta's treatment of the quadrilateral is limited to the cyclic quadrilateral. In fact all his geometry is concerned with figures inscribable in circles, as a survey of his section on *Kṣetragaṇita* will convince anybody.[1] This fact was perhaps vaguely realised by Śrīdhara and Mahāvīra and more clearly by Śrīpati, but Āryabhaṭa II and Bhāskara II seem to have missed the significance of Brahmagupta's theorems completely.

5.5.3. Śrīdhara gives the formula $A=\sqrt{(s-a)(s-b)(s-c)(s-d)}$ for the area of a quadrilateral without leaving any satisfactory clue to enable us to know whether he realised the limitation in its applicability. Even the expression for the area of a trapezium (*T.S.* 42) is given as the formula for the area of the general quadrilateral. Mahāvīra, on the other hand, notes that this expression is not applicable to the *viṣama caturasra* (*G. S. S.* VII. 50), which makes it probable that he was aware of the restriction in the formula $A=\sqrt{(s-a)(s-b)(s-c)(s-d)}$ as well. (It is note-worthy that the examples appended to his *sūtra* embodying this formula, as applicable to the cyclic quadrilateral and the triangle (*G.S.S.* VII. 50), all refer to triangles only, not to

[1]This aspect of Brahmagupta's geometry has been dealt with by the writer in a paper on "The Cyclic Quadrilateral in Indian Mathematics" presented at the 21st session of the *All India Oriental Conference*, 1961.

quadrilaterals). Brahmagupta's expression for the diagonals of a cyclic quadrilateral is repeated in

क्षितिहतौ विपरीतभुजौ मुखगुणभुजमिश्रितौ गुणच्छेदौ ।
छेदगुणौ प्रतिभुजयोः संवर्गयुतेः पदं कर्णौ ॥

(*G. S. S.* VII. 54)

(The flanks multiplied by the base and respectively combined with the opposite flank sides as multiplied by the face are the multiplier and divisor and the divisor and multiplier (respectively) of the sum of the products of the opposite sides. The roots of the quantities so obtained will be the diagonals).

Here too, the stanza does not make it clear that it refers to cyclic quadrilaterals only. The last example occuring under this *sūtra* and intended to cover the case of the *viṣama* quadrilateral gives 195, 260, 125 and 300 for the four sides, i.e. the traditional 39, 52, 25 and 60 of the Brahmagupta quadrilateral multiplied by 5. The treatment of the rational cyclic quadrilateral also follows Brahmagupta's. The area of such quadrilaterals is half the product of the diagonals फलं श्रुतिगुणार्धम् (*G.S.S.* VIII. 107½) since the diagonals cut at right angles.

Mahāvīra's formula for the circum-diameter of a cyclic quadrilateral is neither of restricted application nor vague like Brahmagupta's.

श्रुतिरवलम्बकभक्ता पार्श्वभुजघ्ना चतुर्भुजे त्रिभुजे ।
भुजघातो लम्बहृतो भवेद्बहिर्वृत्तविष्कम्भः ॥

(*G.S.S.* VII. 213)

(The diagonal divided by the altitude and multiplied by the flank is the diameter of the escribed circle in the case of a quadrilateral; that of the triangle is the product of the two sides divided by the altitude.)

Brahmagupta gives this formula for the circum-radius of a trapezium, though it is capable of being applied to any cyclic quadrilateral.

5.5.4. Āryabhaṭa II does not seem to be aware of the existence of cyclic quadrilaterals at all. He has no expression for the circum-radius of a triangle or a trapezium, not to speak of a *viṣama* quadrilateral. He accepts Brahmagupta's formula for the area of a triangle, but rejects the corresponding one for the quadrilateral. His break with the circle tradition in the geometry of rectilinear figures is absolute.

5.5.5. But Śrīpati, between the non-conformists Āryabhaṭa II and Bhāskara II, is a follower of the circle tradition. For him Brahmagupta's formula for the area viz.

$$A=\sqrt{(s-a)\ (s-b)\ (s-c)\ (s-d)}$$

gives the correct area in quadrilaterals as well as triangles[1] and he calculates the circum-radius of quadrilaterals.

चतुर्भुजेष्टश्रुतिपार्श्वबाहुवधस्य चार्द्धं खलु लम्बभक्तम् ।
अतुल्यबाहोः प्रतिबाहुभाग (बाहु ?) वर्गैक्य-मूलस्य दलं हि यद्वा ॥ (*Si. Se.* p. 86)

(In a quadrilateral, half the product of any given diagonal and the flank at its side divided by the altitude is the circum-radius (*hṛdayarajju*). In quadrilaterals with unequal sides it can, alternatively, be half the square root of the sum of the squares of opposite sides).

For forming the rational cyclic quadrilateral, 'Śrīpati', like Brahmagupta, directs that the longest side shall be the base and the shortest the face. Hence the alternative formula.

For the diagonal of a cyclic quadrilateral, he gives Brahmagupta's formula[2]. Like Brahmagupta again he ignores non-cyclic quadrilaterals.

5.5.6. Bhāskara II, as already mentioned, seems to be guided solely by Āryabhaṭa II, in his treatment of Brahmagupta's theorems and results for the cyclic quadrilateral. But he does not cut himself away from them entirely. He reproduces Brahmagupta's theorem for the area of a quadrilateral, adding that the area so obtained will be gross.[3] The theorem

[1] *Si. Se.*, p. 85.
[2] *Si. Se.* p. 86.
[3] It is interesting that the Āryabhaṭa School, which was equally devoted to Bhāskara, reads this verse (*Līl.* 169) with an emendation as

सर्वदोर्युतिदलं चतुःस्थितं बाहुभिर्विरहितं च तद्वधात् ।
मूलमत्र नियतश्रुतौ फलं त्यस्रबाहुजमपि स्फुटं भवेत् ॥
(*Y.B.* p. 257)

The latter half in the published texts of the *Līlāvatī* has मूलमस्फुटफलं चतुर्भुजे स्पष्टमेवमुदितं त्रिबाहुके Brahmagupta's quadrilaterals are *niyataśruti* (with determinate diagonals) because all the possible arrangements of the sides can have only three calculable diagonal-lengths.

about the diagonals also finds a place, though he is at pains to show that the diagonals of a quadrilateral whose four sides only are given are not determinate. Another criticism he has to make is that the calculation is unnecessarily complicated, when the diagonals can be more easily got from Brahmagupta's own method for forming the rational quadrilateral.

अभीष्ठजात्यद्वयबाहुकोटय: परस्परं कर्णहता भुजा इति ।
चतुर्भुजं यद्विषमं प्रकल्पितं श्रुतिस्तु तत्र त्रिभुजद्वयात्ततः ॥
बाह्वोर्वधः कोटिवधेन युक् स्यादेका श्रुतिः कोटिभुजावधैक्यम् ।
अन्या लघौ सत्यपि साधनेऽस्मिन् पूर्वैः कृतं यद् गुरु तन्न विद्मः ॥

(*Līl.* 189-90)

(In the quadrilateral formed according to the rule *abhīṣṭajātya-dvaya......*' the diagonals can be got from the triangles themselves. The product of the *bhujas* combined with the product of the *koṭis* will be one diagonal and the other will be the sum of the reciprocal products of the *bhujas* and *koṭis*). When this easy method is available, I do not know why the earlier mathematicians should have adopted a more difficult method i.e. if a, b, c; x, y, z are the two *jātyas* (right triangles of reference) from which the Brahmagupta quadrilateral is formed, the diagonals will be $ax+by$ and $ay+bx$. This is indeed an easier method for finding the mutually perpendicular diagonals of the Brahmagupta quadrilateral. But it is only in positions where the longest and shortest side face each other that the diagonals will be mutually perpendicular. For the other positions Bhāskara has a supplement in his notes

अथ यदि पार्श्वभुजमुखयोः व्यत्ययं कृत्वा न्यस्तं क्षेत्रं तदा जात्यद्वयकर्णयोर्वधः द्वितीयकर्णः।

(If a flank and the face are interchanged, the product of the hypotenuses of the two *jātyas* is the second diagonal). Thus Bhāskara's method is easier in the case of quadrilaterals formed from known *jātyas*.[1] But Brahmagupta's rule enables us to calculate the diagonals of any cyclic quadrilateral. This Bhāskara

[1]H.C. Bannerji in his notes on "Colebrooke's translation of the Līlāvatī" remarks that Gaṇeśa points out as a fault that Bhāskara's rule requires sagacity in selecting the *jātyas*. What Gaṇeśa actually does is to defend Bhāskara saying that Brahmagupta's rule also is meant for such quadrilaterals only, (See *Līl.* p. 183).

did not realise nor did he realise the cyclic character of Brahmagupta's quadrilateral, which no serious student can easily miss. He ignores the calculations of the circum-radius undertaken by his predecessors.

5.5.7a. Nārāyaṇa Paṇḍita takes the study of the cyclic quadrilateral much farther than Brahmagupta. In fact, though we have not as yet, any clear evidence that he belonged to the Āryabhaṭa School, the advances in circle geometry (and series mathematics) achieved in this school seem to be foreshadowed by the *Gaṇita Kaumudī*. The theorem of the three diagonals is the most important of these.

सर्वचतुर्बाहूनां मुखस्य परिवर्तने यदा विहिते ।
कर्णस्तदा तृतीय: पर इति कर्णत्रयं भवति ।।

(*G. K. Ks. Vya*. 48)

(When the top side and the flank side of any four-sided figure are interchanged, we get a third diagonal called *para*. Thus there are three diagonals).

In a square and a trapezium with three sides equal these three diagonals are equal. In the isosceles trapezium and the rectangle, two diagonals are equal. The half verse dealing with the diagonals of a *viṣama* quadrilateral i.e. कर्णत्रयं समं स्यात् विषमे च चतुर्भुजे नियतम् । seems to be corrupt. The discovery of other manuscripts may help to fix the reading. But it is obvious this interchange of sides without changing the area is possible only in the cyclic quadrilateral, not the general quadrilateral. Hence the statement is equivalent to : three and only three diagonals are possible for four side-lengths of a cyclic quadrilateral.[2]

5.5.7.b The area of a cyclic quadrilateral is then given in terms of these three diagonals.

द्विगुणव्यास विभक्ते त्रिकर्णघातेऽथवा गणितम् ।

(*G. K. Ks. Vya*. 52)

Or, when the product of the three diagonals is divided by twice the circum-diameter, we get the area)

[1]*G.K. Ks. Vya*. 50.

[2]Bhāskara II also seems to have been aware of this fact vaguely.

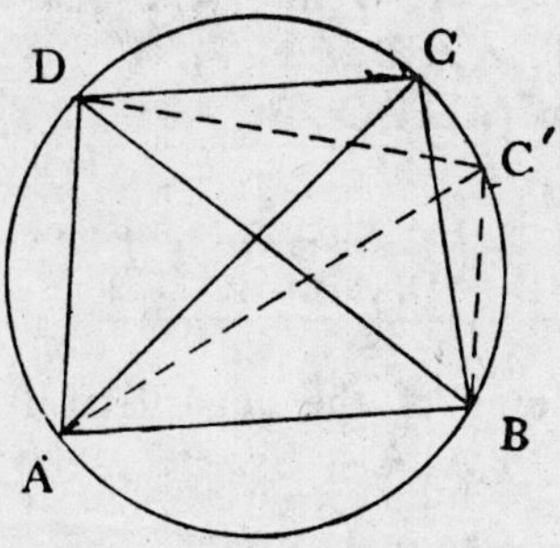

Fig. 6

For, the area of the quadrilateral $ABCD = \triangle ACD + \triangle ACB$

$$= \frac{AC.\ AD.\ CD}{4\,r} + \frac{AC.\ BC.\ AB}{4\,r}$$

(Whether r is the circum-radius)

$$= \frac{AC}{4r}.\ (BC'.\ AD + DC'.\ AB)$$

(Where C′ is the vertex got by interchanging the sides DC and BC)

Now, by Ptolemy's theorem, the sum of the products of the opposite sides of a cyclic quadrilateral = the product of its diagonals.

$$\text{Hence } BC'.\ AD + DC'.\ AB = AC'.\ BD$$

$$\therefore \text{Quadrilateral } ABCD = \frac{AC.\ AC'.\ BD}{4\,r}$$

$$= \frac{AC.\ AC'.\ BD}{2.\ d}$$

Ptolemy's theorem, a knowledge of which is necessary for this derivation, was first proved in India, according to G.R. Kaye,[1] by Brahmagupta's commentator Pṛthūdakasvāmin. The Āryabhaṭa school also has preserved a proof.

Two more expressions are given for the area of a cyclic quadrilateral. The first is

कर्णाश्रितभुजवधयुतिगणिते त(ऽन्य?)स्मिन् श्रवस्यापि विभक्ते ।
चतुराहतहृदयेन द्विसमादिचतुर्भुजे गणितम् ॥

(*G. K. Ks. Vya.*, 134)

(When the diagonal is multiplied by the sum of the products of the sides about the other diagonal and divided by four times the circum-radius, that will be the area of isosceles trapezia and other (cyclic quadrilaterals.)

[1]*Indian Mathematics* (1915), p. 22.

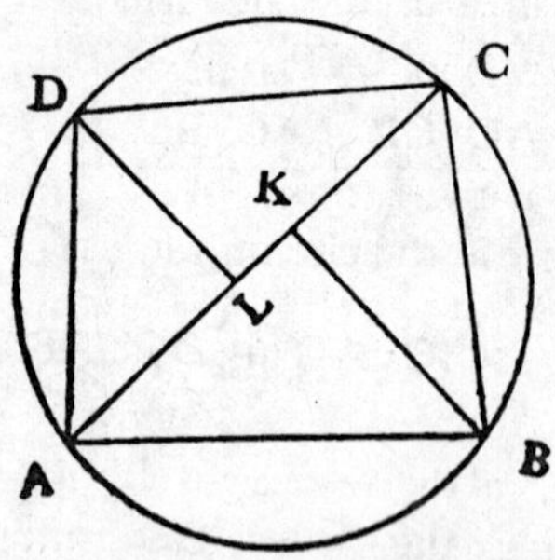

Fig. 7

i.e. the area A of the cyclic quadrilateral ABCD (fig. 7)

$$=\frac{AC(AD.DC+BC.AB)}{4\,r}$$

(where r is the circum-radius or

$$=\frac{BD\,(AD.\,AB+BC.DC)}{4\,r}$$

For, the area of the quadrilateral ABCD $=\triangle ADC+\triangle ABC$

$$=\frac{AC}{2}.\,DL+\frac{AC}{2}\cdot BK$$

$$=\frac{AC}{2}\cdot\frac{AD.DC}{2r}+\frac{AC}{2}\quad\frac{AB.BC}{2r}$$

$$=\frac{AC\,(AD.DC+AB.BC)}{4r}$$

5.5-7c. A minor result about quadrilaterals having equal areas, circum-radii and diagonals is that, if the sides of the first quadrilateral are known, the sides of the second are given by the square roots of the differences between the squares of twice the radius and the squares of the sides of the first quadrilateral taken one by one (*G. K. Ks. Vya.* 135), i.e. if a, b, c, d, r are the sides and circum-radius of the first quadrilateral, the sides of the second are $\sqrt{4\,r^2-a^2}$, $\sqrt{4\,r^2-b^2}$, $\sqrt{4r^2-c^2}$ and $\sqrt{4\,r^2-d^2}$ (for, if the circum-centre O is joined to the vertices of the quadrilateral, the quadrilateral is divided into 4 isosceles triangles and the altitudes of these triangles are $\frac{\sqrt{4r^2-a^2}}{4}$, $\frac{\sqrt{4r^2-b^2}}{4}$, $\frac{\sqrt{4r^2-c^2}}{4}$,

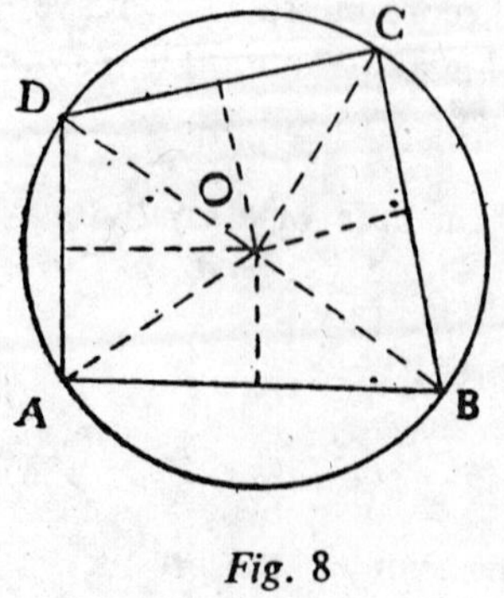

Fig. 8

$$\frac{\sqrt{4r^2-d^2}}{4}$$

Any other quadrilateral in the same circle with the same area can only have these altitudes and half bases interchanged. Hence the result.)

5.5.7.d Nārāyaṇa gives various expressions for the circum-radius, *abādhās* etc.

(1) The circum-radius[1]

$$= \frac{1}{2}\sqrt{\frac{\text{product of diagonals} \times \text{product of flanks}}{\text{product of altitudes}}}$$

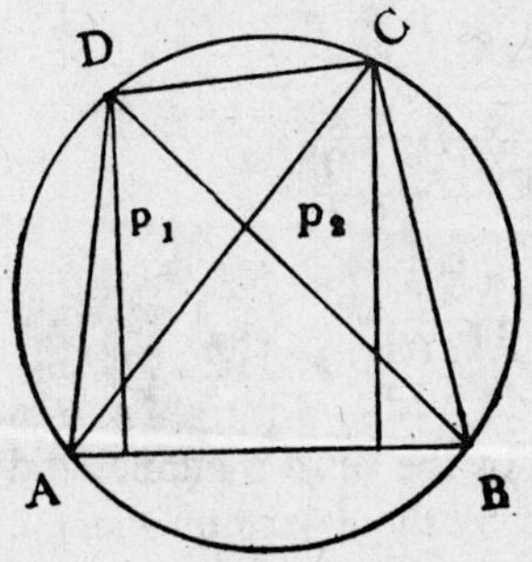

Fig. 9

(For, from $\triangle ADB$,

$$r = \frac{AD \cdot BD}{2p_1}$$

From $\triangle ACB$

$$r = \frac{AC \cdot BC}{2p_2}$$

i.e. $r^2 = \dfrac{AD \cdot BD \cdot AC \cdot BC.}{4\, p_1 \cdot p_2}$

$$r = \frac{1}{2}\sqrt{\frac{AC \cdot BD \cdot AD \cdot BC}{p_1 \cdot p_2}}$$

(2) The circum-radius[2] $= \dfrac{\text{the product of the 3 diagonals}}{4\,A}$

where A is the area of the quadrilateral. This is the converse of the expression for the area.

(3) "The segments of the diagonal are respectively the products of the sides about the diagonal as divided by their sum and multiplied by the other (?) diagonal" *G.K., Ks. Vya.*, 138)[3]

i.e. AE (Fig. 10)

$$= \frac{AD \cdot AB \cdot AC}{AD.AB + BC.CD}$$

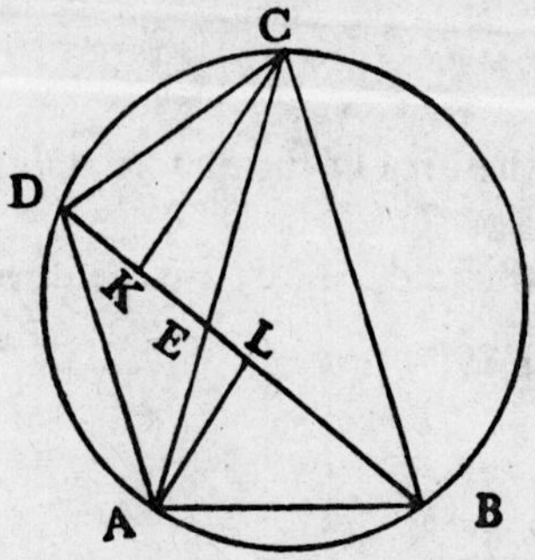

and $CE = \dfrac{BC \cdot CD \cdot AC}{AD.AB + BC.CD}$

$$BE = \frac{AB \cdot BC \cdot BD}{AB.BC + AD \cdot CD}$$

and $DE = \dfrac{AD \cdot CD \cdot BD}{AB.\ BC + AD \cdot CD.}$

Fig. 10

[1] *G.K., Ks. Vya.*, 138.

[2] *Ibid.* 138½.

[3] The word 'other' (*anya*) in अन्यकर्णसंगुणितौ seems to be unnecessary.

(From the similar triangles CEK and AEL

$$\frac{CE}{AE}=\frac{CK}{AL}$$ (CK and AL are the perpendiculars on BD from C and A)

$$\therefore \frac{CE}{AE+CE}=\frac{CE}{AC}=\frac{CK}{CK+AL}$$

$$\therefore CE=\frac{AC \,.\, CK}{CK+AL}=\frac{AC \,.\, \frac{CD.BC}{2r}}{\frac{CD \,.\, BC+AD \,.\, AB}{2r}}$$

$$=\frac{BC \,.\, CD \,.\, AC}{CD \,.\, BC+AD.AB}$$

Similarly the other expressions can be derived).

(4) "The base and the face are divided by the third diagonal and the two flanks are separately multiplied by them (the quotients). The products will be the lower and upper segments of the diagonal".

(*G. K., Ks. Vya*, 139)

i.e. $AE=\frac{AB \,.\, AD}{d_3}$,

$CE=\frac{BC \,.\, CD}{d_3}$ etc., where d_3 is the third diagonal.

(For, as in (3), $CE=\frac{AC \,.\, CK}{CK+AL}$

$$=\frac{AC \,.\, BC \,.\, CD}{(CK+AL)\, 2r}$$

But $\frac{BD}{2}(CK+AL)=A$ (where A is the area of the quadrilateral)

$=\frac{d_1.d_2.d_3}{4r}$ where d_1, d_2, d_3 are the three diagonals.

$$\therefore CK+AL=\frac{2.d_1.d_2.d_3}{BD \,.\, 4r}$$

$$\therefore CE=\frac{AC \,.\, BD \,.\, BC \,.\, CD. \, 2r}{d_1.d_2.d_3. \, 2r}$$

$$=\frac{BC \,.\, CD}{d_3}$$

Similarly for the other segments).

(5) "When twice the area is divided by the product of the base and the diagonal and multiplied by the lower segment of that diagonal, we get the altitude from the tip of the (other) diagonal in isosceles trapezia and other quadrilaterals."

(*G. K., Ks. Vya.*, 140)

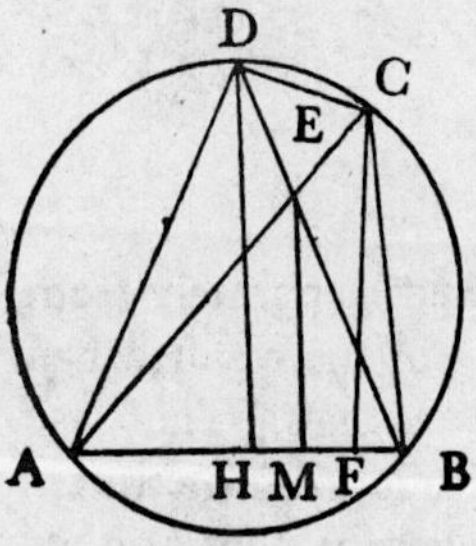

Fig. 11

i.e. $DH \text{ (Fig. 11)} = \frac{AE \,.\, 2\text{ Area}}{AC \,.\, AB}$

and $CF = \frac{BE \,.\, 2\text{ Area}}{BD \,.\, AB}$

(For, from triangle DAB,

$$DH = \frac{AD \,.\, BD}{2r}$$

$$= \frac{AD \,.\, BD}{\frac{d_1 . d_2 . d_3}{2\text{ Area}}}$$

$$= \frac{AB \,.\, AD}{d_3} \times \frac{BD \times 2\text{ Area}}{d_1 . d_2 .\, AB}$$

$$= \frac{AE \times 2\text{ Area}}{AC \,.\, AB}$$

Similarly $CF = \frac{BE \,.\, 2\text{ Area}}{BD \,.\, AB}$)

(6) "The square root of the product of the lower segments of the diagonals as multiplied by the product of the flanks, when divided by twice the circum-radius, becomes the altitude from the point of intersection of the diagonals".

(*G. K., Ks., Vya.*, 142)

i.e. $EM \text{ (Fig. 11)} = \frac{\sqrt{AE \,.\, BE \,.\, AD \,.\, BC}}{2r}$

(From similar triangles DHB and EMB

$$\frac{EM}{DH} = \frac{BE}{BD}$$

or $EM = \frac{BE \,.\, DH}{B.D}$

$$= \frac{BE.\ AD.\ BD.}{2r.\ BD}$$

$$= \frac{BE\ .\ AD}{2r}$$

Similarly, from triangles CAF and EAM,

$$EM = \frac{AE\ .\ BC}{2\ r}$$

or $$EM^2 = \frac{AE.\ BE.\ AD.\ BC}{(2\ r)^2}\Bigg)$$

(7) "The squares of the flanks are subtracted separately from the square of the diameter. The square roots of the results when divided by the respective flanks are called *śakalas*. The base divided by the sum of the *śakalas* is the altitude from the intersection of the diagonals. The *śakalas* multiplied by that altitude are the segments made by it on the base."

(*G. K., Ks. Vya.* 142-142½)[1]

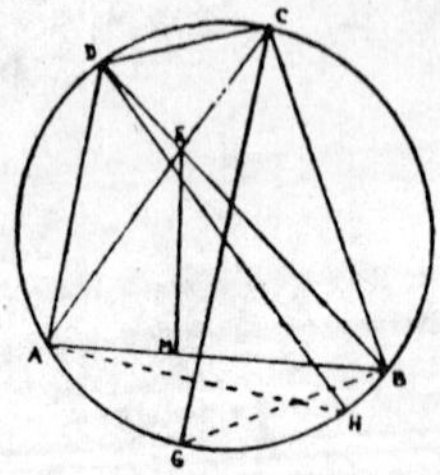

Fig. 12

Let ABCD be a cyclic quadrilateral. CG and DH are diameters through C and D. The diagonals AC, BD intersect at E and EM is the altitude from E.

Then $CG^2 - CB^2 = BG^2$

$DH^2 - AD^2 = AH^2$

$\frac{BG}{BC}$ and $\frac{AH}{AD}$ are given the name of *śakala*.

$$\text{Then } EM = \frac{AB}{\frac{BG}{BC} + \frac{AH}{DA}}$$

$$AM = \frac{BG}{BC} \cdot EM$$

(For, triangles C G B and E A M are similar because

[1]The numbering of the verses is confused in the printed text. The verses run:

बाह्वोः कृती विहीने पृथक् पृथग् व्यासवर्गतो मूले ।
स्वभुजाप्ते शकलाख्ये, तद्युतिहृतभूः श्रवो लम्बः ॥
ते तेन हते शकले श्रुतियुतिलम्बात् कुखण्डे स्तः ।

$\angle CBG = 1$ rt $\angle$ (angle in semi-circle)
$= \angle EMA$

and $\angle CGB = \angle EAM$ (angle in the same segment)

therefore $\frac{EM}{BC} = \frac{AM}{BG}$

$$\therefore AM = \frac{EM.BG}{BC}$$

Similarly $BM = \frac{EM.AH}{AD}$

$$AM + BM = AB = EM\left(\frac{BG}{BC} + \frac{AH}{AD}\right)$$

$$EM = \frac{AB}{\frac{BG}{BC} + \frac{AH}{AD}})$$

We do not know whether this represents the mode of derivation adopted by Nārāyaṇa. If it does, it shows that the theorems

(1) The angle in a semi-circle is a right angle; and

(2) The angles in the same segment are equal;

were known in India in those days.

(8) "If the area is multiplied by the base as divided by half the difference between the squares of the base and the face, we get the altitude of the *sūcī*. The flanks and their *sandhis* (projections on the base) multiplied by the *sūcīlamba* and divided by the altitude at the end of the respective side are the sides and the *ābādhās* of the *sūcī*." (143—143½)

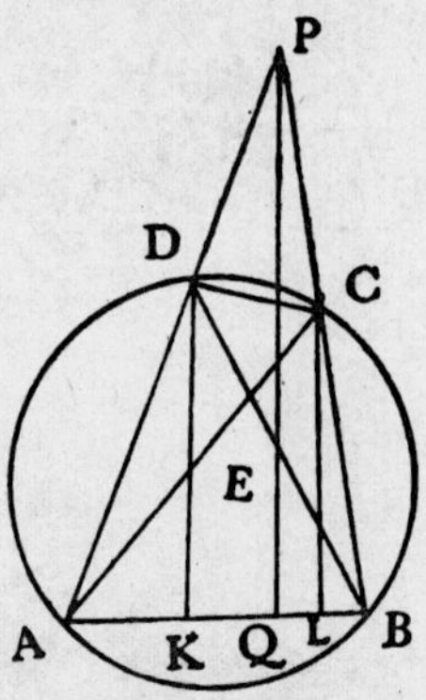

Fig. 13

If PAB is the *sūcīkṣetra* of the cyclic quadrilateral ABCD, the *sūcīlamba*,

$$PQ = \frac{Area . AB}{\frac{AB^2 - DC^2}{2}}$$

$$AQ = \frac{AK. PQ}{DK}$$

$$BQ = \frac{BL . PQ}{CL}$$

$$AP = \frac{AD . PQ}{DK} \; \& \; BP = \frac{BC . PQ}{CL}$$

(These results are derived by Sri Padmakara Dvidedi as follows :

In triangles P D C and P A B

$\angle PDC = \angle ABP$ (since the opposite angles of a cyclic quadrilateral are supplementary)

and $\angle P$ is common

$\therefore$ The two triangles are similar

$$\therefore \frac{AP}{PC} = \frac{AB}{CD} = \frac{BP}{PD}$$

$$\text{i.e. } AP = \frac{PC \cdot AB}{CD} = \frac{AB\,(BP - BC)}{CD}$$

$$BP = \frac{PD \cdot AB}{CD} = \frac{AB\,(AP - AD)}{CD}$$

$$AP + BP = \frac{AB\,(BP - BC + AP - AD)}{CD}$$

$$\text{i.e. } (AP + BP)\,CD = AB\,(AP + BP) - AB\,(BC + AD)$$

$$\text{i.e. } (AP + BP)(AB - CD) = AB\,(BC + AD)$$

$$\text{or } AP + BP = \frac{AB\,(BC + AD)}{AB - CD}$$

$$\text{Similarly } AP - BP = \frac{AB\,(AD - BC)}{AB + CD}$$

Adding 2 AP

$$= \frac{AB\{AB(BC + AD + AD - BC) + CD(BC + AD - AD + BC)\}}{AB^2 - CD^2}$$

$$= \frac{AB\,(2AB \cdot AD + 2CD \cdot BC)}{AB^2 - CD^2}$$

Now, Nārāyaṇa has already given the formula—

$$\text{the area of the cyclic quadrilateral} = \frac{BD(AD.AB + CD.BC)}{4r}$$

$$\therefore AD \cdot AB + CD \cdot BC = \frac{4 \cdot \text{Area} \times r}{BD}$$

$$\therefore AP = \frac{AB \cdot 4 \cdot \text{Area} \cdot r}{BD(AB^2 - CD^2)}$$

Again from similar triangles P A Q & D A K

$$PQ = \frac{AP \cdot DK}{AD}$$

$$=\frac{AB\,.\,4A\,.\,r\,.\,AD\,.\,BD}{BD(AB^2-CD^2)\,.\,2r\,.\,AD}$$

$$=\frac{AB\,.\,Area}{\frac{AB^2-CD^2}{2}}$$

With the help of this the *abādhās* and the sides can be easily calculated.) Thus, if this lemma applies to the general cyclic quadrilateral, it implies the knowledge of the fundamental property of a cyclic quadrilateral, namely that its opposite angles are supplementary.

(9) The last set of calculations in cyclic quadrilaterals given by Nārāyaṇa is rather interesting. He says "the square roots of the differences between the square of the diameter and the square of the diagonal are termed *avakāśa*. The square root of the difference of the square of the diameter and the third diagonal is to be named *guṇa*. The *avakāśas* are multiplied by the *guṇa* and by the diameter. The products are then mutually subtracted and added. The (four) quantities thus obtained when divided by the third diagonal are the differences between the segments of the diagonals. By doing *saṁkramaṇa*[1] with these differences and the whole diagonals (the smaller difference with the smaller diagonal, the bigger with the bigger etc.) we get the segments of the diagonals of the two quadrilaterals".

(*G. K., Ks. Vya.* 145-47)

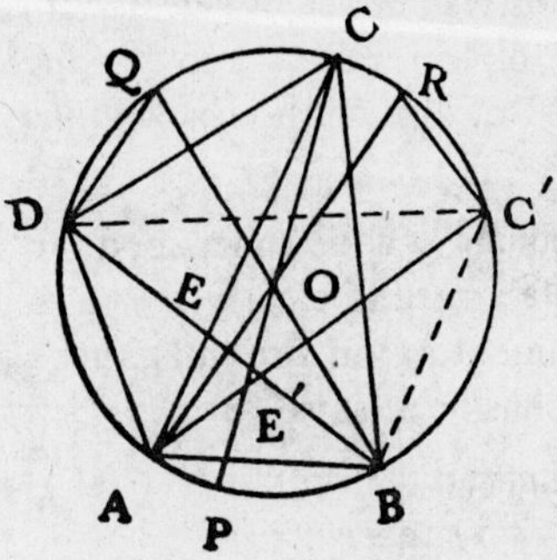

Fig. 14

Let A B C D be a cyclic quadrilateral and ABC′D the quadrilateral in the same circle, obtained by interchanging the sides BC and CD. The diagonals AC, BD, AC′ are joined and the diameters CP, BQ, AR are drawn. Then AP, DQ are *avakāśas* and C′R is the *guṇa*. If d is the diameter, these are respectively equal to

[1]*Saṁkramaṇa* is a technical word for finding two quantities (a & b) from their sum (a+b) and difference (a—b).

$\sqrt{d^2-AC^2}$, $\sqrt{d^2-BD^2}$, $\sqrt{d^2-AC'^2}$. The differences between the segments of the diagonals, namely AE∼CE, BE∼DE, AE′∼C′E′, and DE′∼BE′ are then given by

$$\frac{AP \,.\, C'R \,.\, d \pm DQ \,.\, C'R \,.\, d.}{AC'}$$

and $$\frac{AP \,.\, DQ \,.\, d \pm C'R \,.\, AP \,.\, d}{AC}$$

Then with the help of the known diagonal-lengths, AE, CE, BE, DE, AE′, C′E′, BE′, DE′ can be calculated.

In the next verse we are told how to calculate the areas of the small triangles formed by pairs of these segments with the sides of the quadrilateral.

"When the third diagonal is multiplied by the product of the segments forming that triangle and divided by four times the circum-radius, we get its area". (*Ibid*. 148)

In the notes attached to this verse Nārāyaṇa explains how to get the sides of the quadrilateral formed by sets of these triangles, by c lculating the third side of each triangle from the known area and two sides. These third sides will then be the sides of the quadrilateral.

Thus what Nārāyaṇa does is the inversion of Brahmagupta's method for finding the diagonal-lengths of four side-lengths forming a cyclic quadrilateral. Nārāyāṇa has already shown that three diagonals are possible for any such set of sides. With the three diagonal-lengths and the circum-radius known, the four side-lengths can be calculated. But the method is tedious and complicated.

5.5.8. The mathematicians of the Āryabhata school understood the scope and applicability of Brahmagupta's results clearly.

Parameśvara of the 15th century commenting on Brahmagupta's verse कर्णाश्रितभुजघातैक्यं ...... quoted by Bhāskara, says

यस्मिन् चतुर्भुजे एवं प्रकल्पितं कर्णद्वयं भवति तत्र सर्वदोर्युतिदलमित्यादिना आनीतं क्षेत्रफलं च स्फुटं भवतीति । एतत्क्षेत्रफलान्न्यूनमेव कर्णान्तरकृतक्षेत्रफलानि भवन्ति ।

"In quadrilaterals, the diagonals of which are calculated with the help of this formula, the area obtained by applying the formula *sarvadoryutidala* etc i.e. $A = \sqrt{(s-a)\,(s-b)\,(s-c)\,(s-d)}$

will be the exact area. The areas of quadrilaterals constructed with other diagonals will all be less than this."

The *Līlāvatī* verse अभीष्टजात्य.. .professing to give an easier method for finding the diagonals (than Brahmagupta's कर्णाश्रितभुजघातैक्यं ..) has been criticised[1] as involving the difficult job of selecting two suitable *jātyas* as a preliminary to calculating the diagonals. Parameśvara's explanation of the verse answers this criticism partially.

एतदुक्तं भवति विषमचतुर्भुजस्य चतुर्भिर्भुजैः जात्यद्वयमुत्पाद्य तद्वशात् अभीष्टजात्यद्वयेत्यादिना कर्णत्रयं साध्यमिति । तत्प्रकारस्तु अधिकं भुजं अल्पं भुजं च इष्टराशिना विभजेत् । तत्र लब्धे भुजाकोटी भवतः । ताभ्यां कर्णश्च साध्यः । एवमेकं जात्यक्षेत्रमुत्पन्नं, पुनरस्य कर्णेन चतुर्भुजस्य इतरौ भुजौ विभजेत् तत्र लब्धे अन्यजात्यस्य भुजाकोटी भवतः । ....एवं जात्यद्वंद्वं कृत्वा कर्णा चतुर्भुजे साध्या इति । एवं सिद्धे क्षेत्रे सर्वदोर्युतिदलमिति फलं च संवदति । ....अत्र कोणचतुष्टयस्पृग् वृत्तं च भवति ।

(This is what is said: constructing two *jātyas* with the four sides of the quadrilateral one should calculate the three diagonals with their help. The procedure is this.—The longest and least sides should be divided by a chosen number. The quotients will be the base and the perpendicular side. From them the hypotenuse is to be calculated. Thus one *jātya* is produced. By its hypotenuse the other two sides of the quadrilateral should be divided. The quotients will be the base and the perpendicular sides of the other *jātya*. Thus procuring two *jātyas* one has to calculate the diagonals of the quadrilateral. In the figure obtained in this way the area got by applying the rule सर्वदोर्युति ......agrees with the real area.....In this case there will be a circle passing through the four vertices.)

But the sides will lend themselves to this treament only if they form sides of a Brahmagupta quadrilateral i.e. if the sum of the squares of the longest and shortest sides equals the sum of the squares of the middle ones. Parameśvara or his predecessors in the school have, in addition, investigated the properties of such a quadrilateral, which are mainly

(1) The given sides with the diagonals calculated as above enclose the largest area.

[1]H.C. Colebrooke's translation of the *Līlāvatī* with notes—H.C. Bannerji under V. 190.

(2) Brahmagupta's formula $A=\sqrt{(s-a)(s-b)(s-d)(s-c)}$ is strictly applicable to such quadrilaterals.

(3 A circle can be drawn through the four vertices of such a quadrilateral.

5.5.9. Parameśvara gives a formula for the circumradius of a cyclic quadrilateral after the *Līlāvatī* verse[1] for the *ābādhās* and altitude of a *sūcīkṣetra*

दोष्णां द्वयोर्द्वयोर्धातयुतीनां तिसृणां वधात्।
एकैकोनेतरत्र्यैक्यचतुष्क-वध-भाजिते॥
लब्धमूलेन यद्वृत्तं विष्कम्भार्द्धेन निर्मितम्।
सर्वं चतुर्भुजक्षेत्रं तस्मिन्नेवावतिष्टते॥

("The three sums of the products of the sides taken two at a time are to be multiplied together and divided by the product of the four sums of the sides taken three at a time and diminished by the fourth. If a circle is drawn with the square root of this quantity as radius, the whole quadrilateral will be situated on it.) i.e. the radius

$$=\sqrt{\frac{(ab+cd)(ac+bd)(ad+bc)}{(a+b+c-d)(b+c+d-a)(c+d+a-b)(d+a+b-c)}}$$

if a, b, c, d are the sides. The *Kriyākramakarī*, the anonymous commentary on the *Līlāvatī* belonging to the same school, introduces the same verse with the prelude[2] (अथ चतुष्कोणस्पृशः परिधेः व्यासार्धकल्पनाय करणसूत्रम्) and gives the rationale (*upapatti*) too.

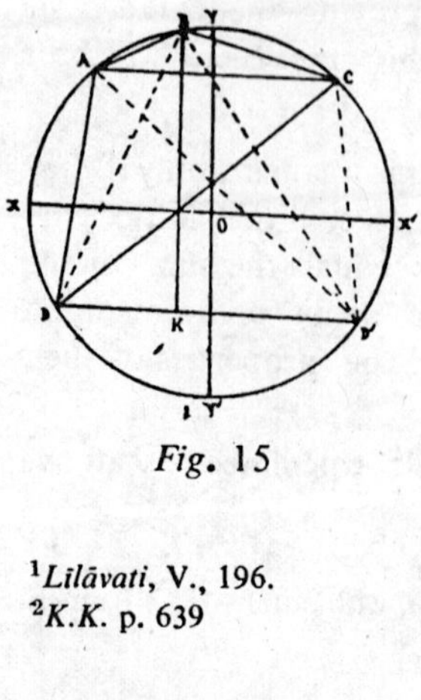

Fig. 15

Draw a circle and two mutually perpendicular diameters XOX′ and YOY′. With AC parallel to XOX′ as *iṣṭakarṇa* (the chosen diagonal) (draw a cyclic quadrilateral ABCD. Join the second diagonal BD. Then

[1]*Līlāvatī*, V., 196.
[2]*K.K.* p. 639

If AD and CD are interchanged AC will still be a diagonal, but we get a third diagonal (*bhāvikarṇa*) different from BD (called *itarakarṇa*). Let D′ be the new position of D. Then D Y′ = D′ Y′

and arc DD′ = arc CD − arc AD

Now if BK is drawn perpendicular to D D′,

$$BK = \frac{BD \cdot BD'}{2r}$$

$$\text{or} \qquad 2r = \frac{BD \cdot BD'}{BK}$$

Again BK will be the sum of the altitudes of the triangles BAC and DAC on AC as base.

$$\therefore \quad \frac{AC \cdot BK}{2} = \text{the area of the quadrilateral.}$$

$$\therefore 2r = \frac{BD \cdot BD'}{BK} = \frac{BD \cdot BD' \cdot AC}{2\ \text{Area}}$$

$$r = \frac{BD \cdot BD' \cdot AC}{4\ \text{Area}}$$

$$= \frac{BD \cdot BD' \cdot AC}{4\sqrt{(s-a)(s-b)(s-c)(s-d)}}$$

$$\text{or} \quad r^2 = \frac{BD^2 \cdot BD'^2 \cdot AC^2}{16(s-a)(s-b)(s-c)(s-d)}$$

$$= \frac{BD^2 \cdot BD'^2 \cdot AC^2}{(a+b+c-d)(b+c+d-a)(c+d+a-b)(d+a+b-c)}$$

$$\text{But} \quad BD^2 = \frac{(bc+ad)(ac+bd)}{ab+cd} \quad \text{(by Brahmagupta's theorem)}$$

$$AC^2 = \frac{(ab+cd)(ac+bd)}{bc+ad}$$

$$BD'^2 = \frac{(cd+ab)(ad+bc)}{ac+bd}$$

$$\therefore BD^2 \cdot BD'^2 \cdot AC^2 = (ab+cd)(ac+bd)(bc+ad)$$

$$\text{or} \quad r = \sqrt{\frac{(ab+cd)(ac+bd)(bc+ad)}{(a+b+c-d)(b+c+d-a)(c+d+a-b)(d+a+b-c)}}$$

This formula was discovered in Europe more than two centuries later i.e. in 1782 by Lhuilier (D. E. Smith. *History of Maths*, Vol. II. 286).

The *Kriyākramakari* also says that the diagonals calculated according to Brahmagupta's rule will be the *niyata-karṇas*, the determinate diagonals, when the quadrilateral is cyclic.

5.5.10. Proof of $\text{Sin}^2 A - \text{Sin}^2 B = \text{Sin}(A+B) \cdot (\text{Sin } A-B)$ and $\text{Sin } A \cdot \text{Sin } B = \text{Sin}^2 \frac{A+B}{2} - \text{Sin}^2 \frac{A-B}{2}$

The *Yuktibhāṣā*, an exposition in Malayāḷam of the mathematical and astronomical knowledge of the time, deals with the cyclic quadrilateral in detail, using it to arrive at trigonometrical results and proving Brahmagupta's results with the help of these. The lemmas first taken up for proof (*Y. B.* pp. 224-227) are

(1) $\text{Sin}^2 A - \text{Sin}^2 B = \text{Sin}(A+B). \text{Sin}(A-B)$

(2) $\text{Sin } A. \text{ Sin } B = \text{Sin}^2 \frac{A+B}{2} - \text{Sin}^2 \frac{A-B}{2}$

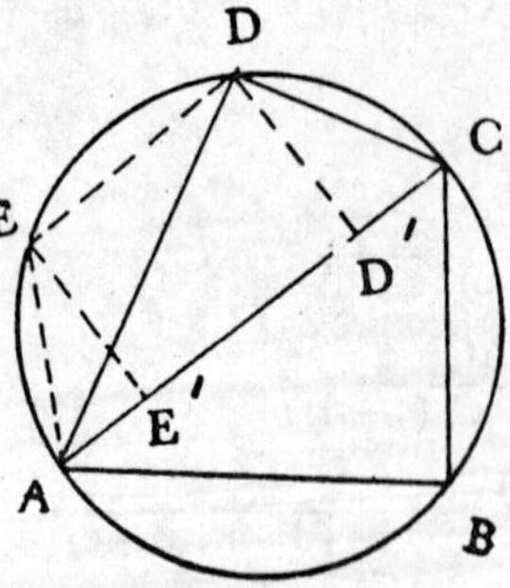

Fig. 16

Let ABCD be a cyclic quadrilateral in which AB > CD,
Along the arc AD set off AE = CD.
Join AE, ED and drop the perpendiculars EE′ and DD′ on AC.
Then arc ED = arc AD − arc DC,
and since arc AE = arc CD,
EE′ = DD′
∴ ED = E′D′ and AD = CE
Now in the triangle DAC, E′ D′ = the difference between the projections of the sides on the base = chord of arc (AD − DC)

AC = the sum of the projections
= chord of arc (AD+DC)

Hence $AD^2 - DC^2 = AD'^2 - D'C^2 = (AD' + D'C)(AD' - D'C)$
= chord of (AD+DC). chord of (AD−DC)...(1)

which, as the editors of the *Yuktibhāṣā* point out, is equivalent to

$\text{Sin}^2 A - \text{Sin}^2 B = \text{Sin}(A+B). \text{Sin}(A-B)$

Also arc AC + arc DE = arc AD + arc CD + arc AD − arc CD
= 2 arc AD

$$\text{arc AC}-\text{arc DE} = \text{arc AD}+\text{arc CD}-\text{arc AD}+\text{arc CD}$$
$$=2 \text{ arc CD}$$

If we put arc AC=A and arc DE=B,

$$\text{arc AD} = \frac{A+B}{2}, \text{ and arc CD}=\frac{A-B}{2}$$

Substituting these values in (1)

$$\text{Sin}^2\frac{A+B}{2} - \text{Sin}^2\frac{A-B}{2} = \text{Sin A. Sin B.}$$

5.5.11. *Derivation of Brahmagupta's expressions for the diagonals of a cyclic quadrilateral*

These results are used to get expressions for the diagonals of a cyclic quadrilateral and its area, and incidentally for proving Ptolemy's theorem (*Y. B.* 228-237)

Let ABCD be a cyclic quadrilateral. Let AB>BC>AD>DC. In arc AD, set of AD′=arc CD and in arc AB set off AB′=BC. Join the middle points X and X′ of the arcs DD′ and BB′.

$$\text{Then arc X A X}' = XD' + D'A + AB' + B'X'$$
$$= XD + CD + BC + BX'$$
$$= \text{arc X C X}'$$

∴ XX′ is a diameter. ∴ arc X′A=arc X′C=arc BC+arc BX′

Similarly arc XA = arc XC=arc CD + arc DX.

Now ch. AD. ch. CD.

$$=\text{ch}^2\frac{AD+CD}{2}$$
$$-\text{ch}^2\frac{AD-CD}{2}$$
$$= AX^2-XD^2$$

$$AB.BC=\text{ch}^2\frac{AB+BC}{2}$$
$$-\text{ch}^2\frac{AB-BC}{2}$$
$$=AX'^2-X'B^2$$

∴ AB. BC+AD, CD

$$=AX'^2+AX^2-X'B^2-XD^2$$

But $AX'^2+AX^2=XX'^2$

(∟XAX′ is the angle in a semi-circle)

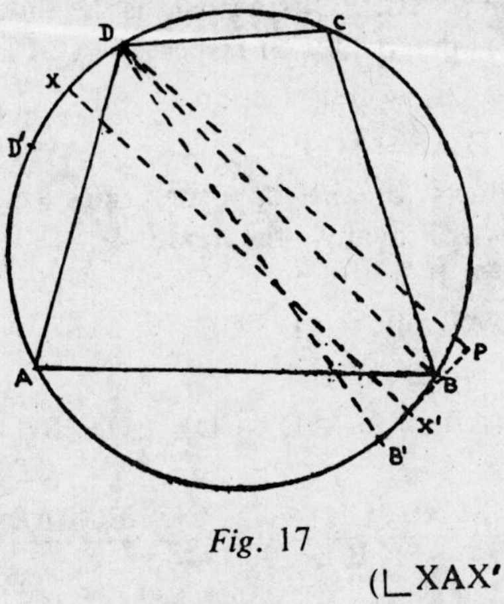

Fig. 17

$$\therefore AB \cdot BC + AD \cdot CD = XX'^2 - XD^2 - X'B^2$$
$$= X'D^2 - X'B^2$$
$$= B'D \cdot BD$$

Therefore the sum of the products of the sides about the first diagonal = the first diagonal × the third.

In the $\triangle$ DBB′, altitude $DP = \frac{B'D \cdot BD}{2r}$ (r is the radius of the circle)

But DP is equal to the sum of the altitudes on AC from B and D

∴ The area of the quadrilateral ABCD

$$= \triangle DAC + \triangle BAC$$
$$= \frac{AC}{2} \times \text{sum of the altitudes}$$
$$= \frac{AC}{2} \cdot DP$$
$$= \frac{AC}{2} \cdot \frac{B'D \cdot BD}{2r}$$
$$= \frac{AC \cdot BD \cdot B'D}{4r}$$

Thus here, Nārāyaṇa's formula for the area of a cyclic quadrilateral is derived and proved incidentally.

Now considering the cyclic quadrilateral AB′CD

$AB' \cdot AD + CD \cdot B'C = AC \cdot BD$ (since BD is the third diagonal with respect to AB′CD

But $AB' = BC$ and $B'C = AB$ by construction,

$$\therefore BC.AD + CD.AB = AC.BD$$

That is, the sum of the products of the opposite sides = the product of the diagonals. This is Ptolemy's theorem.

It has been already proved that

$$B'D \cdot BD = AB \cdot BC + AD \cdot CD$$

and $\quad AC \cdot BD = AD \cdot BC + AB \cdot CD$

$\therefore \quad BD^2 \cdot B'D \cdot AC = (AB \cdot BC + AD \cdot CD)\ (AD \cdot BC + AB \cdot CD)$

$$\text{or } BD^2 = \frac{(AD \cdot BC + AB \cdot CD)\,(AD \cdot CD + BC \cdot AB)}{B'D \cdot AC}$$

$$\text{But } B'D \,.\, AC = AB' \,.\, CD + AD \,.\, B'C$$
$$= BC \,.\, CD + AB \,.\, AD$$
$$\therefore BD^2 = \frac{(AD \,.\, BC + AB \,.\, CD)\,(AD \,.\, CD + BC \,.\, AB)}{AD \,.\, AB + CD \,.\, BC}$$

Hence Brahmagupta's formula.

5.5.12 *Proof of Sin (A±B)=Sin A Cos B±Cos A Sin B*

Next the *Yuktibhāṣā* proceeds to prove (pp. 237-239) geometrically the trigonometrical identity.

$$\text{Sin }(A \pm B) = \text{Sin A. Cos B} \pm \text{Cos A. Sin B},$$

which is found stated without explanation or proof in the section on *Jyotpatti* appended to Bhāskara II's *Siddhāntaśiromaṇi*. The statement of this lemma, famous in the Āryabhaṭa school as the जीवे परस्परन्याय, is found in many works of this school including the *Tantrasaṃgraha* and it is unanimously attributed to Mādhava of Saṃgamagrāma (14th century).

जीवे परस्परनिजेतरमौर्विकाभ्यामभ्यस्य विस्तृतिदलेन विभज्यमाने ।
अन्योन्ययोगविरहानुगुणे भवेतां । . . . . ॥

(*T. S.* II. 16½)

"(The sine-chords of two arcs reciprocally multiplied by the other (cosine) chord and divided by the radius, when added or subtracted from each other will be the sine-chords of the sum or difference of the arcs). In the quadrant XOY, let PX = A and PQ = 2B. Draw PR perpendicular to OX. Join OP. Let P′ be the middle point of PQ. Join OP′ cutting chord PQ in S. Join SR. Then PR and PS are the *bhujajyās* (sine-chords) of arcs A and B respectively, and OR and OS are the corresponding *koṭijyās* (cosine chords). Now OSPR is a cyclic quadrilateral.

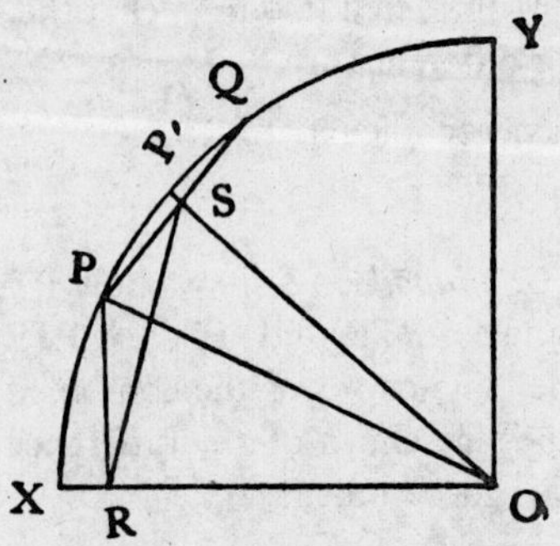

Fig. 18

$$\therefore PR \,.\, OS + PS \,.\, OR = OP \,.\, SR$$

But SR is the sine-chord of the combined arc X P′ i.e. of (A+B)

$$\therefore SR = r. \text{ Sin } (A+B) = \frac{PR.OS + PS.OR}{OP}$$ (Where r is the radius of the circle)

$$= \frac{r \text{ Sin A. } r \text{ Cos B} + r \text{ Sin B. } r \text{ Cos A}}{r}$$

$$\text{or Sin } (A+B) = \text{Sin A. Cos B} + \text{Cos A. Sin B.}$$

That RS is the sine-chord of the sum of the arcs X P and P P′ is proved by the *Yuktibhāṣā* editors as follows. Let the circle through R, P & S be drawn. Then RP, PS are whole chords and RS is the whole chord of arc (RP+PS). In a circle with twice the radius all these will be half-chords or sine-chords

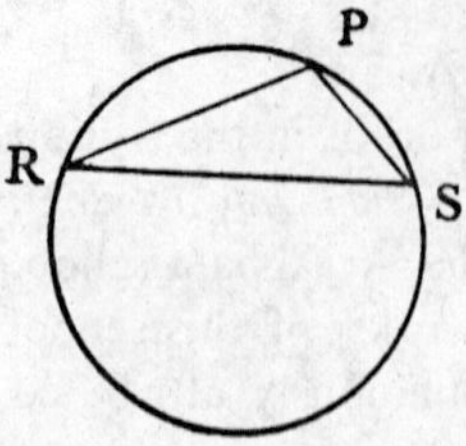

∴ RS is the sine-chord of arc (RP+PS) i.e. of arc (X P+P P′), if P R and P S are the sine-chords of arc XP and PP′.

(*Y. B.* p. 214)

In the last foot of the above verse, an alternative method is given for finding the sine of the sum and difference arcs :

. . . . यद्वा स्वलम्बकृतिभेदपदीकृते द्वे ।

("Or the roots of the (two) differences of the squares of the sine and its altitude when added to or subtracted from each other will be the sine-chord of the sum or difference arc.")

That is, if, in Fig. 18, PT is drawn perpendicular to RS,

$$RS \text{ is } \sqrt{PR^2 - PT^2} + \sqrt{PS^2 - PT^2}$$

Nīlakaṇṭha, commenting on *A.B. Gaṇitapāda*. 12 shows how PT can be found from the similar triangles PTR and PSO or PTS and PRO (*A.B.* p. 87). Hence the sine-chord of the combined arc can be easily found out. For the sine-chord of the difference arc the difference of the two roots is to be taken.

5.5.13. *Proof of area of a cyclic quadrilateral*

$$= \sqrt{(s-a)(s-b)(s-c)(s-d)}$$

Lastly the expression for the area of a cyclic quadrilateral in terms of its sides is derived. (*Y. B.* pp. 247-257)

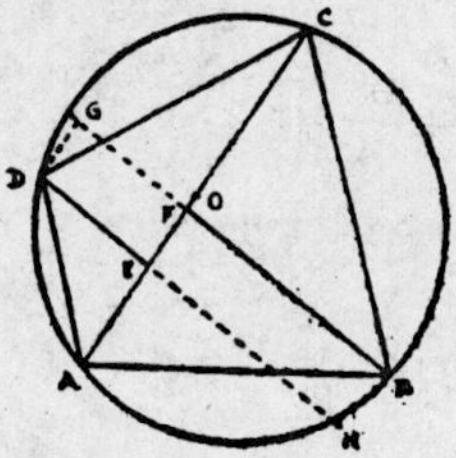

Fig. 19

Let A B C D be a cyclic quadrilateral. Join the diagonal AC and draw the altitudes D E and B F on it. If D E and B F are produced and D G and B H are drawn perpendicular to them, D G B H forms a rectangle with one side = the sum of the altitudes and the other side = the distance between the feet of the altitudes i. e., E F. Let O be the middle point of A C.

Then in $\triangle$ DAC

E O = half the difference between the projections of the sides on A C

$$= \frac{1}{2} \cdot \frac{CD^2 \sim AD^2}{AC}$$

Similarly from $\triangle$ BAC, $FO = \frac{1}{2} \cdot \frac{AB^2 \sim BC^2}{AC}$

$$\therefore EF = EO + FO = \frac{1}{2\,AC} \left\{ (AB^2 + CD^2) - (AD^2 + BC^2) \right\}$$

(If the altitudes fall on the same side of O, EF will be EO$\sim$FO; but still the final expression will be the same)

$$\because HD^2 = BD^2 - BH^2$$

$$= BD^2 - EF^2$$

$$= BD^2 - \left\{ \frac{(AB^2 + CD^2) - (AD^2 + BC^2)}{2\,AC} \right\}^2$$

Now the area, A, of the quadrilateral A B C D

$$= \triangle DAG + \triangle BAC = \frac{1}{2} AC. \text{ sum of altitudes}$$

$$\therefore A^2 = \frac{AC^2}{4} \left[BD^2 - \left\{ \frac{(a^2 + c^2) - (b^2 + d^2)}{2\,AC} \right\}^2 \right]$$

$$= \frac{AC^2}{4} \cdot \frac{4AC^2 . BD^2 - \{ (a^2 + c^2) - (b^2 + d^2) \}^2}{4\,AC^2}$$

$$= \left(\frac{AC. BD}{2} \right)^2 - \left\{ \frac{(a^2 + c^2) - (b^2 + d^2)}{4} \right\}^2$$

$$= \left(\frac{ac + bd}{2} \right)^2 - \left\{ \left(\frac{a^2}{4} + \frac{c^2}{4} \right) - \left(\frac{b^2}{4} + \frac{d^2}{4} \right) \right\}^2$$

by Ptolemy's theorem

$$= \left(\frac{ac}{2}+\frac{bd}{2}+\frac{a^2}{4}+\frac{c^2}{4}-\frac{b^2}{4}-\frac{d^2}{4}\right)\left\{\frac{ac}{2}+\frac{bd}{2}-\frac{a^2}{4}-\frac{c^2}{4}+\frac{b^2}{4}+\frac{d^2}{4}\right\}$$

$$=\left\{\left(\frac{a}{2}+\frac{c}{2}\right)^2-\left(\frac{b}{2}-\frac{d}{2}\right)^2\right\}\left\{\left(\frac{b}{2}+\frac{d}{2}\right)^2-\left(\frac{a}{2}-\frac{c}{2}\right)^2\right\}$$

$$=\left(\frac{a}{2}+\frac{c}{2}+\frac{b}{2}-\frac{d}{2}\right)\left(\frac{a}{2}+\frac{c}{2}-\frac{b}{2}+\frac{d}{2}\right)$$

$$\left(\frac{b}{2}+\frac{d}{2}+\frac{a}{2}-\frac{c}{2}\right)\left(\frac{b}{2}+\frac{d}{2}-\frac{a}{2}+\frac{c}{2}\right)$$

$$= (s-d)\,(s-b)\,(s-c)\,(s-a),$$

Where $s = \frac{a}{2}+\frac{b}{2}+\frac{c}{2}+\frac{d}{2}$

$$\therefore A = \sqrt{(s-a)\,(s-b)\,(s-c)\,(s-d)}$$

The occurrence of these geometrico-algebraical proofs in the *Yuktibhāṣā* and *Kriyākramakarī* shows conclusively that, as in any country making any real progress in sciences, in India too proofs were required and esteemed.

CHAPTER VI

THE TRIANGLE

6.1. The earliest mention of a triangle in Indian literature is, as has been already seen, perhaps in the Vedas; as *trirasri* in the *Ṛgveda* and *tribhuja* in the *Atharvaveda*. The *Śulbasūtras* deal with one particular kind of triangle only, viz. the isosceles triangle with base angles equal to $\tan^{-1} 2$. The name *prauga* given to this triangle is entirely ungeometrical with no reference to any of its geometrical properties. The word means the forepart of the shaft of a chariot or cart. Indirectly the *Śulbasūtras* were quite familiar with another important type of triangle, the right angled triangle. But such triangles were viewed not as triangles but as halves of squares and rectangles cut by their diagonals.

Nothing much of the properties of the *prauga* was investigated. From the construction for a *prauga* of given area we can infer that the authors of the *Śulbasūtras* knew that the area of the triangle $= \frac{1}{2}$ the area of a rectangle on the same base and with the same altitude $= \frac{1}{2}$ the base x altitude. Similarly from the construction of the *ardhyās* or half bricks by cutting a rectangular brick along a diagonal, it is clear that they knew that the area of a right triangle also is half the product of the perpendicular sides $= \frac{1}{2}$ base $\times$ altitude.

Another property of the isosceles triangle known and used by the *Śulbasūtras* is that the line joining its vertex to the middle point of the base is perpendicular to the base.

The early Jainas had no use for the triangle. Most of the ancient works do not mention the triangle at all. The *Prajñāpanopāṅgam* and the *Bhagavatī Sūtra*, speaking about the arrangement of atoms (or shots) say that one of the arrangements can be in the form of a triangle.

6.2. Āryabhaṭa's rules for the area of a triangle is

त्रिभुजस्य फलशरीरं समदलकोटीभुजार्धसंवर्गः

(*A.B. Gaṇitapāda* 6)

(The area of a triangle is the product of half its side (base) and the altitude.)

It is difficult to decide what the word *samadalakoṭī* here means. Bhāskara I and Parameśvara take the word to mean *avalambaka* (altitude), the perpendicular side common to the two triangles formed by the altitude. Nīlakaṇṭha, on the other hand, explains it as the perpendicular which divides the base into two equal parts and adds that Āryabhaṭa is speaking about the equilateral triangle useful in dividing the circle into 6 parts, though the formula is equally applicable to the general triangle. But Bhāskara being much nearer to Āryabhaṭa in time, his opinion should be considered weightier.

To calculate the *ūrdhvabhujā* (the height of the tetrahedron) referred to in the second half of the same verse, Nīlakaṇṭha makes use of verse 8 which reads :

आयामगुणे पार्श्वे तद्योगहृते स्वापातरेखे ते ।
विस्तरयोगार्धगुणे ज्ञेयं क्षेत्रफलमायामे ॥

and interprets the *pāta* as the orthocentre and he shows why all the three perpendicular bisector altitudes have necessarily to be concurrent.[1]

6.3. In Brahmagupta's hands, the triangle gets fuller treatment, but curiously enough, for him the triangle[2] is a quadrilateral with one side = 0. Thus he says :

स्थूलफलं त्रिचतुर्भुजबाहुप्रतिबाहुयोगदलघातः ।
भुजयोगार्धचतुष्टयभुजोनघातात् पदं सूक्ष्मम् ॥

(*Br. Sp. Si XII.* 21)

(The gross area of a three or four-sided figure is the product of half the sums of the opposite sides. The exact area is got as

[1]We have no means of knowing how much of the mathematical knowledge found in the commentaries belongs to Āryabhaṭa's time and how much is later. The very obscurity of Āryabhaṭa's statements will indicate that his codified rules were meant to be reinforced by mueh oral instruction.

[2]G.R. Kaye (The Source of Hindu Mathematics. *J.R.A.S.* 1910 p. 753) says "Brahmagupta gives the area of the cyclic quadrilateral as

$$\sqrt{(s-a)(s-b)(s-c)(s-d)}$$

which is an extension of the well-known theorem of Heron for triangles", suggesting thereby that the Indians got the formula from the Greeks. But it is more probable that the Indians discovered the theorem for the cyclic quadrilateral first and extended it to cover the triangle, and then as a proof for it, when applied to the triangle, showed how the usual expression $\frac{1}{2}$ base x altitude could be equated to this. Hence the occurrence of the expression for the area of a triangle in Heron can hardly be advanced as an argument for India's indebtedness to him.

the square root from the product of four sets of half the sum of the sides each respectively diminished by one side.) That is, if a, b, c, d, are the sides of a quadrilateral the gross area $=\frac{a+c}{2}\cdot\frac{b+d}{2}$. When one of the sides is zero the quadrilateral becomes a triangle, whence the ***sthūla*** area of a triangle

$$=\frac{\text{base}}{2}\cdot\frac{\text{sum of the sides}}{2}=\frac{a}{2}\cdot\frac{b+c}{2}$$

The second part of the verse says the exact area of a triangle $=\sqrt{s(s-a)(s-b)(s-c)}$ (where s is the semi-perimeter,) a formula generally credited to Heron of Alexandria.[1]

Brahmagupta does not give the derivations or proofs of his formulae or theorems. But the derivation of this formula was known in the Āryabhaṭa school at least by the 15th century, and possibly much earlier.

The next verse gives formulae for calculating the *ābādhās*, the segments of the base made by the altitude (or rather the projections of the sides on the base) and the altitude.

भुजकृत्यन्तरहृतहीनयुता भूर्द्विभाजिताबाधे ।
स्वाबाधावर्गोनात् भुजवर्गान्मूलमवलम्ब: ॥

(*Br. Sp.-Si.* XII 22)

(The base diminished by and combined with the difference of the squares of the sides divided by the base, and then divided by two gives the *ābādhās*. The square root of the square of the side diminished by the square of its *ābādhā* is the altitude.)

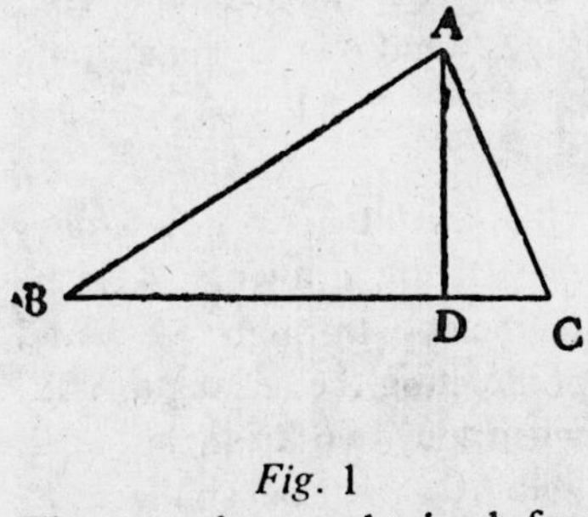

Fig. 1

In fig. 1 AD is the altitude and BD & CD are the *ābādhās*.

$$BD \text{ or } CD=\tfrac{1}{2}\left(a\pm\frac{b^2\sim c^2}{a}\right)$$

$$\text{and } AD=\sqrt{AB^2-BD^2} \text{ or } \sqrt{AC^2-CD^2}$$

These results are obtained from a consideration of the right triangles ABD & ACD in the *Yuktibhāṣā* (p. 222).

[1]T. Heath—*A History of Greek Mathematics* Vol. II p. 103.

For, from these,

$$BD^2 = c^2 - AD^2$$
$$CD^2 = b^2 - AD^2$$
$$BD^2 - CD^2 = c^2 - b^2$$
$$\text{i.e. } (BD+CD)(BD-CD) = c^2 - b^2$$
$$a.(BD-CD) = c^2 - b^2$$
$$\text{or } BD - CD = \frac{c^2 - b^2}{a}$$
$$\text{or } 2BD - a = \frac{c^2 - b^2}{a}$$
$$BD = \frac{a + \frac{c^2 - b^2}{a}}{2}$$
$$\text{and } a - 2CD = \frac{c^2 - b^2}{a}$$
$$\text{or } CD = \frac{a - \frac{c^2 - b^2}{a}}{2}$$

Then $AD^2 = \text{side}^2 - (\text{its } \textit{ābādhā})^2$

XII., 27 gives an expression for the circum-radius of a triangle.

त्रिभुजस्य वधो भुजयोर्द्विगुणितलम्बोद्धृतो हृदय-रज्जुः ।
सा द्विगुणा त्रिचतुर्भुजकोणस्पृग्वृत्तविष्कम्भः ॥

(The product of the sides divided by twice the altitude is the circum-radius, and twice that is the diameter of the circle which touches the vertices of the triangle and the quadrilateral)

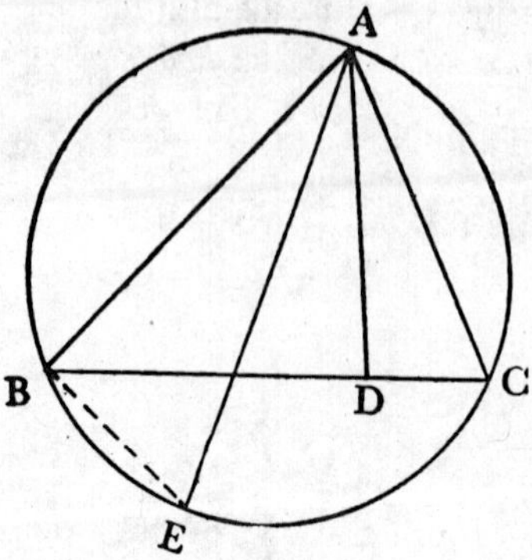

Fig. 2

i.e. If r stands for the circu-radius of the triangle ABC (fig. 2) and AD is its altitude

$$r = \frac{AB \,.\, AC.}{2 \,.\, AD}$$

This result is now-a-days proved by drawing AE the diameter through A and completing the triangle ABE, when the two triangles ABE and ADC will be similar ($\llcorner$ ABE $=$ $\llcorner$ ADC (rt $\llcorner$s) and $\llcorner$ AEB $=$ $\llcorner$ACB ($\llcorner$s in the same segment.)

$$\therefore \frac{AB}{AE} = \frac{AD}{AC}$$

$$\text{or } AE = 2r = \frac{AB \; . \; AC}{AD}$$

But the method adopted by Brahmagupta might have been more akin to the proof given in the *Yuktibhāṣā* for the theorem that the product of two contiguous chords divided by the diameter is equal to the altitude dropped from the meeting point of the chords on the line joining their ends.[1] For, arcs took the place of angles in Indian astronomical geometry. The *Y.B.* proof is as follows. Let ABC be a triangle inscribed in its circum-circle and AD its altitude. Through E the middle point of the arc AC, let the diameter EF be drawn. Let FG be parallel to BC meeting the circumference in G. EG is joined and EH is drawn parallel to FG.

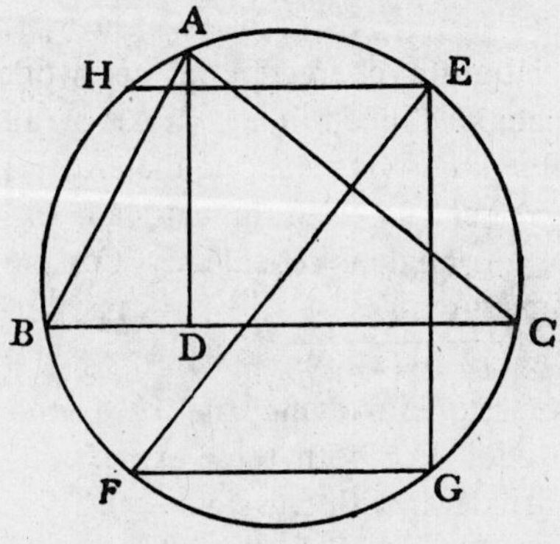

Fig. 3

$$\text{Then } GF = EH = \text{chord of arc } EH$$

$$\text{But arc } EH = \text{arc } BAC - 2 \text{ arc } CE$$

$$= \text{arc } BAC - \text{arc } AC$$

$$= \text{arc } AB.$$

$$\therefore \text{Chord } EH = GF = \text{chord } AB$$

Then from the similar triangles A C D and E F G,

$$\frac{EF}{FG} = \frac{AC}{AD}$$

$$\text{or } EF = 2r = \frac{AC \, . \, FG}{AD} = \frac{AC \, . \, AB}{AD}$$

$$\therefore r = \frac{AC \, . \, AB}{2 \, AD}$$

[1] *Y.B.* pp. 244-246.

Similar triangles are not dealt with separately. But in the computations connected with the quadrilateral such triangles play an important part.

6.4. Śrīdhara too treats the triangle as a particular case of the quadrilateral. The expressions[1] for the area of a triangle are

$$A = \sqrt{s(s-a)(s-b)(s-c)}$$
$$\text{and } A = \tfrac{1}{2} \text{ base x altitude.}$$

Śrīdhara does not give any rule for the calculation of the altitude (*lamba*) and the *ābādhās*.

6.5. In keeping with the elaborate character of his work Mahāvīra's treatment of the triangle is much fuller. First of all he classifies triangles into three kinds (1) *sama*, equilateral (2) *dvisama*, isosceles and (3) *viṣama*, scalene.[2] Brahmagupta'a expression for the gross area of a triangle is repeated.[3] For the exact area, in addition to Brahmagupta's

$$A = \sqrt{s(s-a)(s-b)(s-c)},$$

Mahāvīra gives the familiar expression involving the base and altitude.[4] i.e. $A = \frac{1}{2}$ base x altitude. For both these expressions the triangle is treated as a quadrilateral with one side $= 0$.

The expressions for the altitude and *ābādhās* of a triangle occur in verse 49, ch. VII.

Mahāvīra indulges in a few calculations from the gross and exact areas of geometrical figures, which appear like a mere display of computational ingenuity, but, nevertheless, show that he was acquainted with the special expressions for the area of equilateral triangles. The calculation for the equilateral triangle is :

धनवर्गान्तरमूलं यत्तन्मूलाद्द्विसंगुणितम् ।
बाहुस्त्रिसमत्रिभुजे समस्य वृत्तस्य विष्कम्भः ॥

(*G.S.S.* VII. 168½)

(The square root of the square root of the difference of the squares of the (two) areas multiplied by two is the side in an equilateral triangle and the diameter of a circle.)

[1] *T.S.* 43.
[2] *G.S.S.* VII 4-5.
[3] Ibid. VII. 7.
[4] Ibid VII. 50.

For, the gross area of an equilateral triangle, A_a.

$$= \text{product of half the sums of the opposite sides}$$

$$= \frac{a}{2} \cdot a = \frac{a^2}{2}$$

The exact area $A_e = \frac{\sqrt{3}}{4} \cdot a^2$

$$\therefore \; A_a^2 - A_e^2 = \frac{a^4}{4} - \frac{3a^4}{16} = \frac{a^4}{16}$$

$\therefore$ The 4th root of $A_a^2 - A_e^2 = \frac{a}{2}$

$$\text{or } a = 2\left(A_a^2 - A_e^2\right)^{\frac{1}{4}}$$

The calculation of the base and side of an isoceles triangle similarly reveals the knowledge that the altitude of an isosceles triangle bisects the base.

फलवर्गान्तरमूलं द्विगुणं भून्यविहारिकं बाहुः ।
भूम्यर्धमूलभक्ते द्विसमत्रिभुजस्य करणमिदम् ॥

(*G.S.S.* VII. 171½)

(Twice the square root of the difference between the squares of the areas is the base, and the approximate area is the side, if these are divided by the square root of half the base. This is the calculation in an isosceles triangle) i.e, if A_a and A_e are the areas,

$$\text{base} = \frac{2\left(A_a^2 \sim A_e^2\right)^{\frac{1}{2}}}{\left(A_a^2 \sim A_e^2\right)^{\frac{1}{4}}} = 2\left(A_a^2 \sim A_e^2\right)^{\frac{1}{4}}$$

$$\text{and side} = \frac{A_a}{\left(A_a^2 \sim A_e^2\right)^{\frac{1}{4}}}$$

For, if a is the base and b the side

$$A_a = \frac{ab}{2} \text{ and } A_e = \frac{a}{2}\left(b^2 - \frac{a^2}{4}\right)^{\frac{1}{2}}$$

$$\therefore \quad A_e^2 = \frac{a^2}{4}\left(b^2 - \frac{a^2}{4}\right) = \frac{a^2b^2}{4} - \frac{a^4}{16} = A_a^2 - \frac{a^4}{16}$$

$$\frac{a^4}{16} = A_a^2 - A_e^2$$

$$\therefore \quad a = 2\left(A_a^2 - A_e^2\right)^{\frac{1}{4}}$$

and $$b = \frac{A_a}{\frac{a}{2}} = \frac{A_a^{1}}{\left(A_a^2 - A_e^2\right)^{\frac{1}{4}}}$$

In VII, 213 is given the expression for the circum-diameter equivalent to Brahmagupta's :

$$r = \frac{\text{product of the sides}}{2 \times \text{altitude}}$$

But Mahāvīra is the first to speak of a circle inscribed in a triangle and of its diameter. Says Mahāvīra

परिधेः पादेन भजेदनायतक्षेत्रसूक्ष्मगणितं तत् ।
क्षेत्राभ्यन्तरवृत्ते विष्कम्भोऽयं विनिर्दिष्टः ॥

(*G.S.S.* VII 223½)

("The exact area of any (rectilinear) figure other than a rectangle should be divided by one fourth the perimeter. This is specified as the diameter of the inscribed circle.)

The rule is quite general. The rationale is clear from the attached figures.

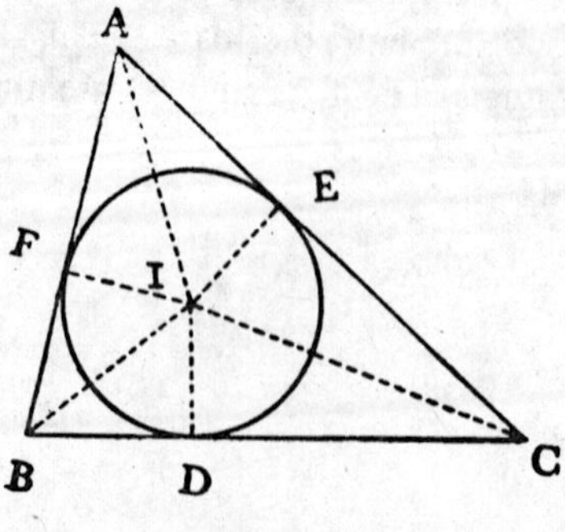

Fig. 4

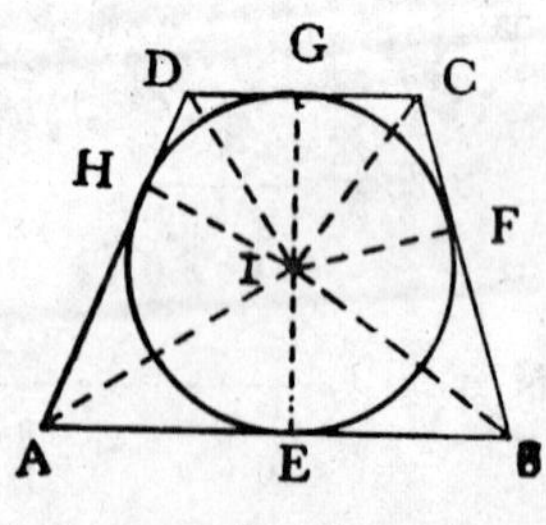

Fig. 5

For, if the in-centre, I is joined to the vertices of the figure and to the points of contact of the circle with the sides, the whole figure is divided into as many triangles as there are sides. These have the sides of the figure as bases, the lines joining I to the vertices as sides and the in-radii at the points of contact as the altitudes.

Hence the area of the triangle

$$= \frac{a \cdot r}{2} + \frac{b \cdot r}{2} + \frac{c \cdot r}{2}$$ (Where a, b, c are the lengths of the sides and r the in-radius)

$$= \frac{r}{2}(a+b+c)$$

$$\therefore \ r = \frac{\text{area}}{\frac{a+b+c}{2}} = \frac{\text{area}}{\frac{1}{2}\text{ perimeter}}$$

The expression gives us no clue which will enable us to know whether Mahāvīra knew that the in-centre is on the bisectors of the angles of the figure.[1]

6.6 Āryabhaṭa II and Śrīpati give the expressions for the exact area of a triangle[2] and the *ābādhās* and the altitude.[3] Śrīpati gives the expression for the circum-radius too, but Āryabhaṭa omits it. Both these authors investigate the conditions under which a triangle or a quadrilateral is possible. According to Āryabhaṭa

शुष्यति कश्चिद्यदि दोरखिलं भुजयोगखण्डकतः ।
शुद्धे बाहोर्योगजखण्डे क्षेत्रं न तद्भवति ॥

(*Ma. Si.* XIV. 64)

(If every side is subtractable from the semi-perimeter, the recti-linear figure is possible. If the semi-perimeter is subtractable from any of the sides, that is not a closed figure.) This, though the wording is different, means the same as 'no side is to be greater than or equal to the sum of the remaining sides'. We do not know whether any proof other than visual demonstration with rods (which Bhāskara II advocates) was sought for or found. (That the shortest distance between any two points is the straight line joining the two points would have been the proof to suggest itself.) Śrīpati gives the same condition, but in

[1]It may be pointed out here that Heron's derivation of the formula $A=\sqrt{s(s-a)(s-b)(s-c)}$ (T. Heath, *History of Greek Mathematics* Vol. II. p. 320) employs the triangles formed by joining the in-centre to the vertices of a triangle.

[2]*Ma. Si.* XIV. 69 and 78 & *Si. Se.* p. 85.

[3]*Ma. Si.* XIV 66-67 & *Si. Se.* p. 85.

a more straight-forward way and without mentioning the triangle.

चतुर्भुजायामखिलस्य वा स्यादवक्रबाहोरधिका(त्)भुजाच्चेत् ।
ऊनस्समो वेतरबाहुयोगो ज्ञेयं तदक्षेत्रमुदारधीभिः ॥

(*Si. Se.* XIII, 26, 27)

(If in a quadrilateral with straight sides the sum of the other sides is less or equal to the greatest side, the wise should understand that it is not a closed figure.)

Bhāṣkara II repeats this condition. Another new feature in Āryabhaṭa's treatment of the triangle is that he is aware that the altitude of a triangle may sometimes fall outside the triangle.

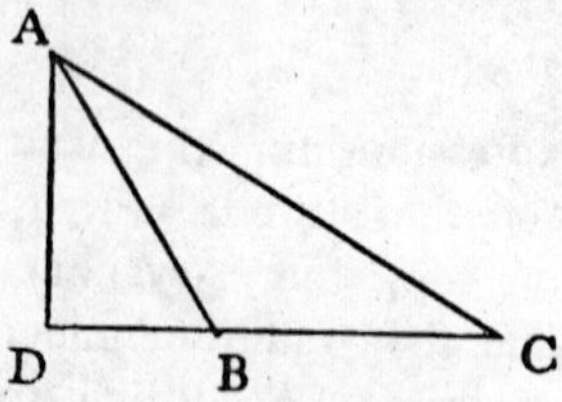

Fig. 6

This happens in an obtuse-angled triangle and then the projection or *ābādhā* of the smaller side about the obtuse angle lies outside the triangle.

लघुबाहोराबाधा व्यस्ता यदि सा बहिर्भवेत् क्षेत्रात् ।

(*Ma. Si.* XIV. 67)

(The *ābādhā* of the shorter side is *vyasta* if it lies outside the figure.) *Vyasta* being the opposite of *samasta* (combined) and so meaning separated, diminished, it should be interpreted here to mean 'negative'. Most likely Āryabhaṭa is the inspirer of Bhāskara's *ṛṇābādhā* (negative projection) also.

6.7. Bhāskara II has nothing new to add on the triangle. On the other hand he even omits to notice that the triangle can have a circumscribing circle. Bhāskara's achievements were in the circle and the sphere.

6.8. Nārāyaṇa Paṇḍita's treatment of *kṣetragaṇitam* is as elaborate as that of Mahāvīra. All the knowledge of triangles contained in his predecessors' works is found in the *Gaṇitakaumudī*, often in a more precise form. Besides, he has new expressions for the circum-radius and area of a triangle.

अबधावधेन हीनो लम्बकवर्गोऽवलम्बकविभक्तः ।
तत्कृतिभूकृतियोगान्मूलदलं जायते हृदयम् ॥

(*G. K. Ks. Vya.*, 133)

(The square of the altitude diminished by the product of the *ābādhās* is divided by the altitude. Half the square root of the sum of the square of this and the square of the base is the circum-radius.)

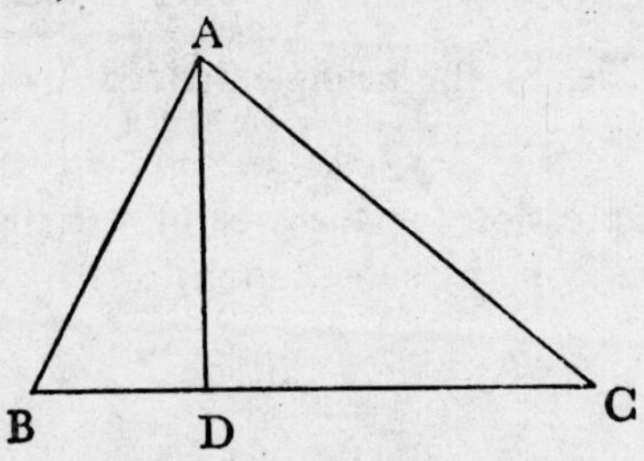

Fig. 7

Let ABC be a triangle and AD its altitude. Then BD, CD are the *ābādhās* and the rule says that the circum-radius

$$= \tfrac{1}{2}\sqrt{BC^2 + \left(\frac{AD^2 - BD \,.\, DC}{AD}\right)^2}$$

For,[1] by the already known formula for the circum-radius,

$$r = \frac{AB \,.\, AC}{2\,AD}$$

$$r^2 = \frac{AB^2 \,.\, AC^2}{4\,AD^2}$$

$$= \frac{(AD^2+BD^2)\,(AD^2+CD^2)}{4\,AD^2}$$

$$= \frac{AD^4+AD^2\,(BD^2+CD^2)+BD^2 \,.\, CD^2}{4\,AD^2}$$

$$= \frac{AD^4+AD^2\,(BC^2-2BD \,.\, CD)+BD^2 \,.\, CD^2}{4\,AD^2}$$

$$= \frac{AD^2 \,.\, BC^2+AD^4-2AD^2 \,.\, BD \,.\, CD+BD^2 \,.\, CD^2}{4\,AD^2}$$

$$= \tfrac{1}{4}\left\{BC^2 + \left(\frac{AD^2-BD \,.\, CD}{AD}\right)^2\right\}$$

$$\therefore r = \tfrac{1}{2} \,.\, \sqrt{BC^2 + \left(\frac{AD^2-BD \,.\, CD}{AD}\right)^2}$$

[1]The editor of the *G.K.* gives this derivation. The other derivation given by him involving equality of angles is unlikely to have been the one adopted by the author.

Nārāyaṇa's new expression for the area of a triangle is embodied in

चतुराहतहृदयहतं त्रिभुजभुजानां वधं गणितम् ।

(*G. K., Ks. Vya.*, 134)

(The product of the sides of the triangle divided by four times the circum-radius is its area.)

This well-known expression for the area of a triangle is easily obtained from the basic one by substitution)

$$A = \frac{a}{2} \cdot \text{altitude}$$

$$= \frac{a}{2} \cdot \frac{bc}{2r} = \frac{abc}{4r}$$

If the altitude is produced to meet the circum-circle, the portion beneath the base can be calculated with the help of the *sūtra* :

लम्बहृदवधाघातो वृत्तस्पर्शी भवेदधोलम्बः ।
अवधे मिथो भुजघ्न्यौ लम्बाप्ते तद् भुजौ स्याताम् ॥

(*G. K. Ks. Vya.*, 101)

(The lower part of the altitude which touches the circum-circle is the product of the *ābādhās* of the base divided by the altitude. The *ābādhās* reciprocally multiplied by their sides and divided by the altitude are the sides.)

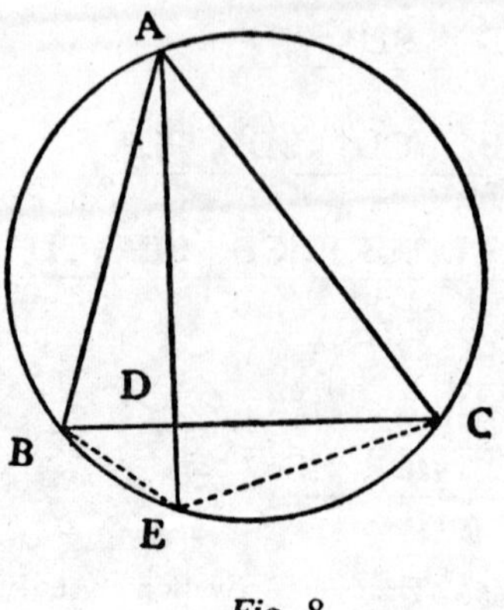

Fig. 8

That is, if, in the $\triangle$ ABC inscribed in a circle, AD the altitude is produced to E on the circum-circumference

$$DE = \frac{BD.\ DC}{AD}$$

$$BE = \frac{BD.\ AC}{AD}$$

and $$CE = \frac{CD.\ AB}{AD}$$

These results can be easily derived from the similar triangles ADC, BDE, and ADB, CDE.

6-9.1. For the mathematics of the next period dominated by Saṃgamagrâma Mādhava, which may be termed the period of the later Āryabhaṭa school and extends roughly from the middle of the 14th century to the 17th century A.D., we have to turn to the *Yuktibhāṣā* and the *Kriyākramakarī* though the computational formulae arrived at in the school are to be found in many a work like the *Karaṇapaddhati*, the *Tantrasaṃgraha* and the *Sadratnamālā*. In this school, whose motto could very well have been, "Let none but an astronomer enter these portals", mathematics was completely subservient to the needs of astronomy. The geometry of the circle and of chords was its chief study. The triangle and the quadrilateral are now formed not by lines but by chords. The chief merit of the *Yuktibhāṣā* is that it preserves for us the rationales and proofs developed in the school, whereas the other schools either did not have them or did not preserve them.

6.9.2. The theorem that the altitude of an isosceles triangle bisects its base and its corollary that in a scalene triangle the altitude is nearer to the shorter side are stated but taken as axiomatic.[1] Other theorms stated are:[2]

(1) Two right angled triangles are similar if the hypotenuse and one side of one triangle are respectively parallel and perpendicular to one side and the hypotenuse of the other triangle.

(2) Two right triangles are similar if (1) the three sides of the one are parallel to the three sides of the other, (2) the three sides of the one are perpendicular to the three sides of the other.

(3) In similar figures the ratios of corresponding sides are equal.

6.9.3.1. The following propositions are supplied with proofs. (1) The area of a triangle $= \frac{1}{2}$ base $\times$ altitude.[3] The proof is best demonstrated with the triangle placed in such a way that

[1] *Y.B.* p. 144.
[2] *Y.B.* p. 85.
[3] *Y.B.* p. 223.

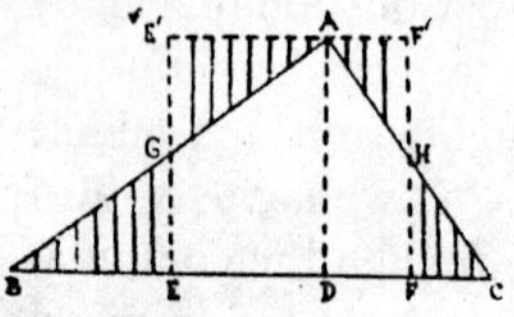

Fig. 9

the longest side forms the base. Let ABC be such a triangle and AD its altitude. The middle points E and F of the *ābādhās* BD and DC are joined to the middle points G and H of AB and AC. The portion EGB is lifted and placed in such a way that BG coincides with AG and triangle EGB occupies the position E′GA. Similarly FCH is placed in the position AF′H. The △ ABC is now transformed into the rectangle EFF′E′ whose sides are the altitude and half the base of the triangle. Hence the area of a triangle = ½ base. altitude. In the same context Brahmagupta's experssions for the *ābādhās* are also derived (the derivation is already given).

(2) The altitude of a triangle is equal to the product of the two sides (other than the base) divided by the diameter of the circum-circle.

The earlier mathematicians gave this inverted as the

$$\text{circum - diameter} = \frac{\text{product of sides}}{\text{altitude}}.$$

The proposition as given above also does not exactly correspond to its statement in this school. The *Yuktibhāṣā* says. "If the product of the sides of a triangle, which are invariably chords of a circle, is divided by the diameter of that circle, the quotient will be the altitude of the triangle whose base will be the chord of the sum of the arcs of the sides."[1] This is an annotatory translation of

ज्ययोः परस्परं घातः त्रिज्याप्तो लम्ब इष्यते ।

(The mutual product of the sine chords divided by the radius is regarded as the altitude) which is found in one version of the *Tantraśaṃgraha* with a Malayālam commentary.[2] The proof is already given in 6.3.

[1] *Y.B.* p. 231.

[2] The Transcript of the Manuscript belonging to Deśamaṅgalath Nampūtiri, p. 53. The *Tantrasaṁgraha* published from the Trivandrum Manuscripts library does not seem to contain the full text. Or, perhaps the *Tantra saṁgraha* has different versions with considerable variations in the volume of contents.

6.9.3.2. *The area of a triangle* $= \sqrt{s(s-a)(s-b)(s-c)}$

Here what is attempted is not a geometrical derivation of the formula but a justification in which by elaborate algebraical reasoning it is shown that $\frac{1}{2}$ base $\times$ altitude

$= \sqrt{s(s-a)(s-b)(s-c)}$. The actual derivation was effected by viewing the triangle as a cyclic quadrilateral with one side $= 0$.

6.9.3.3. *Nilakaṇṭha Somayājin while commenting on Āryabhaṭa I's*

ऊर्ध्वभुजा-तत्संवर्गार्द्धं: स घन: षडश्रिरिति ।

(*A.B.*, *Gaṇitapāda* 6)

dwells at length on how to calculate the *ūrdhvabhujā* (the height of the tetrahedron), which is the perpendicular dropped on the plane of the base from the vertex. Obviously this line is the one joining the apex to the circum-centre-cum-ortho-centre of the basal equilateral triangle. In this connection Nīlakaṇṭha proves that the perpendicular bisectors of the sides of an equilateral triangle are concurrent. "The ends of the base will be equidistant from all the parts of the perpendicular line from the top vertex. For, the middle point of the base is equidistant from its ends and the perpendicular will fall at this point, since the triangle is equilateral. Otherwise the equality of the left and right sides will vanish, in which case the *ābādhās* will be unequal. For demonstrating this, two rods of the length of the base . . . . . . Moreover, the very fact that the *koṭis* of the *dalas* (the right triangles into which the altitude divides the triangle) are equal, proves that their bases also are equal, since the hypotenuses are equal. This is indicated by the word *samadalakoṭi*. Hence in an equilateral triangle, the ends of the base are equidistant from the point of intersection of the altitude and the base. Again the left and right ends of the base will be equidistant from the upper portions of the altitude also, since the altitude goes up vertically and is equally inclined to the base. It is only when one of the ends is inclined that the parts above will be nearer to one end and farther away from the other end. If there is non-inclination (*udāsīnatva*=indifference, sitting above), the distances will be equal. Similarly, the distances of each part of the perpendicular at the middle of the right side from the ends of the right side will be equal. And these two (perpendiculars) have to meet somewhere. For, how can a line starting from the left end of the base reach the middle of the right side without passing the line going vertically up from the middle of the base? Since this line is inclined to the left side, the line from the middle point will be inclined to the line from the end. Hence the intersection of these two will be equidistant from all the three vertices, since every point on the perpendicular to the base will be equidistant from its ends and every point on the perpendicular at the middle of the right side will be equidistant from the ends of the right side. Hence the ends of the base as well as those of the right side will be at the same distance from the point of intersection of the two perpendiculars .... In this way, all the three vertices will be equidistant from the point of intersection. By the same argument, the perpendicular at the centre of the left side also will pass through the intersection" (*A. B. Gaṇitapāda*. pp. 31-32). Though this proof is not quite in the Euclidean tradition, it is ably and logically reasoned out.

6.10. *The Theorem of the Square on the Hypotenuse*

Bhāskara II in his *Vāsanābhāṣya* on *Līlāvatī* asserts in a poetical and philosophical vein, that all mathematics is really *trairāśika*, proportion. Modifying it a bit, we can say that all Indian geometry and trigonometry is really the theorem of the square of the hypotenuse. A pre-occupation with this theorem and the right triangle is a legacy of the *Śulbasūtras* (vide chapter II).

The statement of the theorem in the *Śulbasūtras* has reference to the sides and diagonals of squares and rectangles, whereas Āryabhaṭa's statement is

यश्चैव भुजावर्गः कोटिवर्गश्च कर्णवर्गः सः

(*A. B. Gaṇitapāda*. 17)

(That which is the square of the base (*bhujā*) and that which is the square of the perpendicular (*koṭi*) that is the square of the hypotenuse.)

[1]Under *Līlāvatī*, 239.

If we look into the etymology of the word *bhuja* the significance of this statement is not far to seek. *Bhuja* coming from *bhuj* to bend and retained in its original sense in *bhujaṅga* means a curve, an arc. When astronomy demanded a close study of the circle, the chord stretched across the arc came to be called *bhujajyā* (the bow string of the arc) or *bhuja* for short. When the sine chord i.e. the half-chord of the arc assumed importance in astronomy, *bhujajyā* came to mean the half chord. The remainder of the quadrant and its half-chord, being as it were suspended from the tip of the *bhuja* came to be called *koṭi*.[1] The transition from the *tiryaṅmānī* and *pārśvamānī* of the *Śulbasūtras* to the *bhuja* and *koṭi* is thus reminiscent of a change in the application of the theorem. The construction of the altars and fire-places no more provided scope for the exercise of mathematical talents. The elaborate ritual itself was giving place to another form of worship. At the same time astronomy was gaining in importance. And the theorem of the square of the diagonal, the chief tool in the hands of the altar-builder, again became the chief tool in the hands of the astronomer.[2] Now it is really the equivalent of the trigonometrical identity

$$\text{Sin}^2 A + \text{Cos}^2 A = 1.$$

or

$$(r.\ \text{Sin}\ A)^2 + (r.\ \text{Cos}\ A)^2 = r^2.$$

6.10.1. *Proof of the theorem of the square of the hypotenuse*

In Bhāṣkara's *Bījagaṇita* the rationale for the theorem is given.

दो:कोट्यन्तरवर्गेण द्विघ्नो घातः समन्वितः ।
वर्गयोगसमः स स्याद् द्वयोरव्यक्तयोर्यथा ॥

(*Bījagaṇita*, 129)

(Twice the product of the *bhuja* and *koṭi* combined with the square of their difference will be equal to the sum of their squares, just as it is so for two algebraical quantities.) The

[1] *Koṭi* also means a 'curve' derivatively.

[2] The supreme importance of this theorem and the theory of proportion in astronomy is pointed out by Nīlakaṇṭha Somayājin in his notes on V. 17, in the words

भुजकोटीकर्णन्यायेन त्रैराशिकन्यायेन चोभाभ्यां सकलं ग्रहगणितं व्याप्तम् ।

(*A.B.* Part I. p. 100)

commentators Kṛṣṇa and Gaṇeśa make the method clear. Let ABC be any right triangle. 4 triangles equal and similar to this are put together as in the figure with the hypotenuses forming the sides of a square. Then a small square with sides equal to the difference of the *bhuja* and *koṭi* will be formed at the centre, the rest of the bigger square having been filled up by the four triangles. The area of each triangle $= \frac{1}{2}$ *bhuja* $\times$ *koṭi*

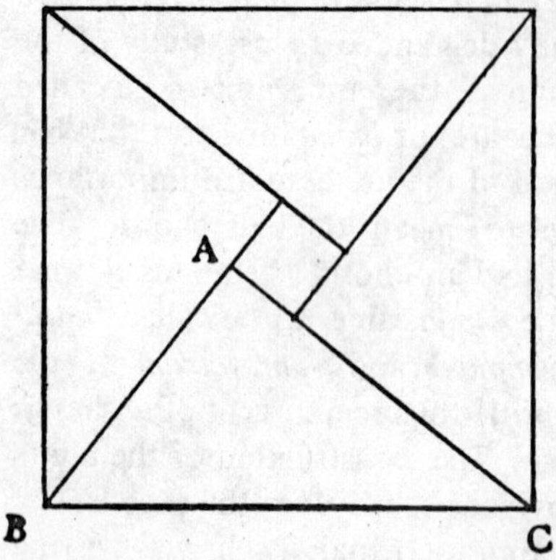

Fig. 10

Hence area of 4 △ s = 2 *bhuja. koṭi*

∴ The bigger square = (*bhuja* ∼ *koṭi*)^2 + 2 *bhuja. koṭi*
= *bhuja*^{2} + *koṭi*^{2}.

This is a geometrico-algebraical proof. A fully geometrical proof also is given by these commentators.[1]

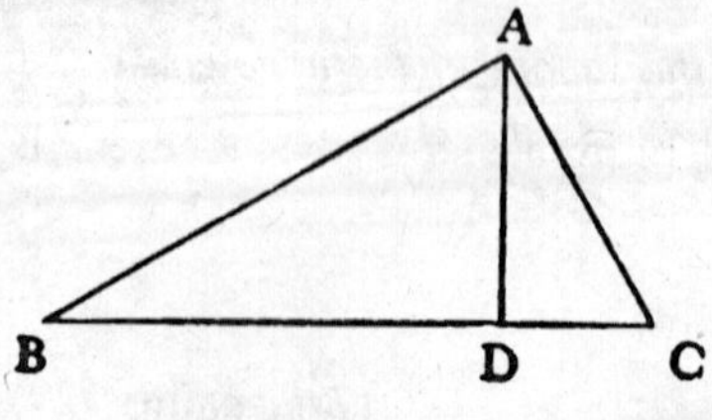

Fig. 11

Let ABC be a triangle right angled at A. Let AD be drawn perpendicular to BC. Then the △ s ADB, ADC and ABC are similar to each other.

∴ from △ s ABD and ABC

$$\frac{AB}{BC} = \frac{BD}{AB} \quad \text{or BD} = \frac{AB^2}{BC}$$

Similarly from △ s ADC and ABC

$$DC = \frac{AC^2}{BC}$$

[1]Colebrooke (*Miscellaneous Essays*. p. 395) says that this proof given by Wallis is in his treatise on angular sections (ch. V) is given by Bhāskara in his *Bījagaṇita*.

$$\therefore \; BD + DC = BC = \frac{AB^2 + AC^2}{BC}$$

$$\text{or } BC^2 = AB^2 + AC^2$$

6.10.2. The proof given in the *Yukti bhāṣā*, i.e. of the Ārya-bhaṭa school is purely demonstrational.[1]

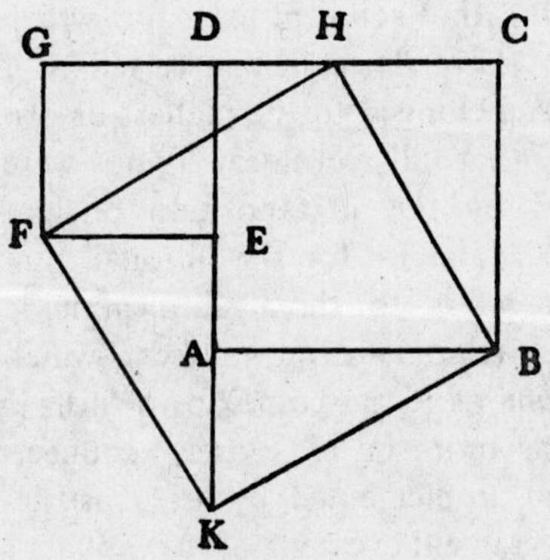

Fig. 12

The square ABCD with side equal to the *bhuja* (a) and the square DEFG with side equal to the *koṭī* (b) are placed side by side, with two sides of each falling in the same line as shown in the figure. From the combined line GC, GH is marked off equal to a. HF is joined and the square HFKB on HF is constructed so as to be over the first two squares.

Then HF is the hypotenuse of the triangle and square HFKB is the square on the hypotenuse. The only parts of the two squares a^2 and b^2 lying outside this, are two right triangles FGH and HCB and these are equal to the $\triangle$s FEK and AKB which lie inside HFKB but outside the other two squares. Hence the square on the hypotenuse = the sum of the squares on the *bhuja* and *koṭi*.

6.11. *Rational rectilinear figures*

Another field in which the theorem of the square of the hypotenuse was extensively used was the formation of rectilinear figures with rational sides. The beginnings of the interest in rational figures are discoverable in the *Śulbasūtras*. Āpastaṁba, according to his commentators, gives a general solution for a right-angled triangle with a given side in his rules for constructing

[1] *Y.B.* p. 72.

right angles on a given side (a) with a cord whose length is divided into $\frac{5}{4}$ a and $\frac{3}{4}$ a or $\frac{5}{12}$ a and $\frac{13}{12}$ a

Brahmagupta gives general rational solutions for the isosceles and scalene triangle, the rectangle, the isosceles trapezium (*dvisama*), the trapezium with three sides equal (*trisama*) and the quadrilateral (*viṣama*). All this is done by judicious juxtaposition of rational right triangles, the technical term for which is *jātyā*. This word also may have its own tale to tell, as if the rational right triangle alone belonged to the highest or original species. It is quite likely that all rectilinear figures were in one sense viewed as formed by the juxtaposition of right triangles. The term used by Mahāvīra for the rational right triangle is *janya*, which perhaps refers to the algebraical mode of formation of the sides of these triangles from numbers, which he calls *bījas*. The *Āryabhaṭīyam* as it has come down to us is a collection of rules and formulae more or less loosely connected and evidently intended to be supplemented by oral instruction and does not help us to gauge the extent of Āryabhaṭa's mathematical knowledge. No section on rational figures is incorporated in it. Still there is reason to believe that he was acquainted with rational right triangles and the method of constructing other rational figures out of them by juxtaposition. The apparently meaningless direction for drawing triangles and quadrilaterals 'त्रिभुजं च चतुर्भुजं च कर्णाभ्याम् ।" (*Gaṇitapāda* 13) becomes intelligible if we remember that these were generally formed out of rational right triangles.

6.11.1.1. Brahmagupta s solution for the rational isosceles triangle is

कृतियुतिरसदृशराश्योर्बाहुर्घातो द्विसंगुणो लम्बः ।
कृत्यन्तरमसदृशयोर्द्विगुणं द्विसमत्रिभुजभूमिः ॥

(*Br. Sp. Si.* XII.33)

(The sum of the squares of two unequal numbers is the side, twice their product the altitude and twice the difference of the squares of the unequal numbers is the base in an isosceles triangle).

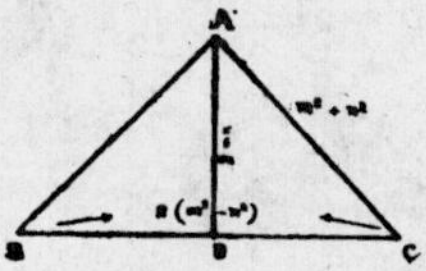

Fig. 13

Manifestly, the triangle is formed by the juxtaposition of two equal rational right triangles of sides $m^2 - n^2$ and $2mn$ and hypotenuse $m^2 + n^2$, the sides $2mn$ being made to coincide with each other.

6.11.1.2 *To get a rational scalene triangle*

इष्टद्वयेन भक्तो द्विघेष्टवर्गः फलेष्टयोगार्धे ।
विषमत्रिभुजस्य भुजाविष्टोनफलार्धयोगो भूः ॥

(*Br. Sp. Si.* XII. 34)

(The square of an optional number is divided by two other optional numbers separately. Halves of the sums of the quotients and the optional number (i.e. the respective divisor) are the sides of the scalene triangle. The base is half the sum of the quotients diminished by the respective (divisor) optional number) i.e. the sides are

$$\tfrac{1}{2}\left(\frac{m^2}{p} + p\right),\ \tfrac{1}{2}\left(\frac{m^2}{q} + q\right) \text{ and } \tfrac{1}{2}\left(\frac{m^2}{p} - p\right) + \tfrac{1}{2}\left(\frac{m^2}{q} - q\right)$$

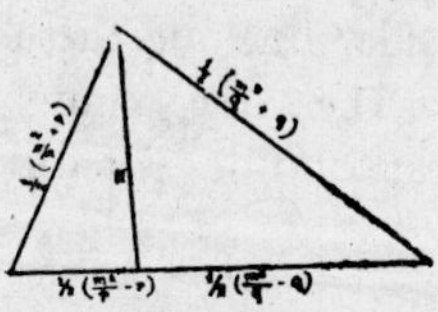

Fig. 14

As Sudhakara Dvivedi explains, the triangle is formed by juxtaposing two right triangles with a common perpendicular side = m, which becomes the altitude in the combined scalene triangle.

The problem then reduces to finding a general solution for the right triangle with one side about the right angle given. This is given in the next verse.

इष्टस्य भुजस्य कृतिर्भक्तोनेष्टेन तद् दलं कोटिः ।
आयतचतुरश्रस्य क्षेत्रस्येष्टाधिकः कर्णः ॥

(*Br. Sp. Si.* XII. 35)

(The square of the given side divided by an optional number, diminished by the same and halved is the perpendicular side, and the same quotient with the optional number added is the diagonal in a rectangular figure)

i.e. the sides of the right triangle are

$$a,\ \tfrac{1}{2}\left(\frac{a^2}{m} + m\right) \text{ and } \tfrac{1}{2}\left(\frac{a^2}{m} - m\right)$$

where a is the given side and m is any arbitrarily chosen number. For if b & c are the other two sides, $c^2 - b^2 = a^2$

$$\text{or } (c - b)(c + b) = a^2$$

$$c + b = \frac{a^2}{c - b} = \frac{a^2}{m} \text{ putting } c - b = m$$

$$\text{Then } c = \tfrac{1}{2}\left(\frac{a^2}{m} + m\right)$$

$$\text{and } b = \tfrac{1}{2}\left(\frac{a^2}{m} - m\right)$$

Hence the solution for the scalene triangle will be

$$\tfrac{1}{2}\left(\frac{m^2}{p} + p\right), \tfrac{1}{2}\left(\frac{m^2}{q} + q\right) \text{ and } \left\{\tfrac{1}{2}\left(\frac{m^2}{p} - p\right) + \tfrac{1}{2}\left(\frac{m^2}{q} - q\right)\right\}$$

In the solution for the right triangle if we put a = n and remove fractions we get the usual general solution of the right triangle, viz. $2\,mn$, $m^2 - n^2$, $m^2 + n^2$.

6 11.1.3. *To construct a rational isosceles trapezium*

Baudhāyana[1] notices that an isosceles trapezium can be made out of two rectangles with one common side, one of these being cut diagonally into two right triangles. The same knowledge is utilised by Brahmagupta in his general solution for the rational isosceles trapezium, which is

आयतकर्णों बाहू भुजकृतिरिष्टेन भाजितेष्टोना ।
द्विहृता कोट्यधिका भूर्मुखमूना द्विसमचतुरश्रे ॥

(*Br. Sp. Si.* XII. 36)

(The lateral sides are the diagonal of the rectangle. The square of the base of the rectangle is divided by an arbitrarily chosen number, then diminished by that number and halved. This is separately combined with and diminished by the perpendicular side of the rectangle. The greater of the two results is the base and the less the face of the isosceles trapezium.)

Here the *bhuja* is chosen to be the altitude (p) of the trapezium. Then the *koṭi* $= \tfrac{1}{2}\left(\frac{p^2}{m} - m\right) = k$ where m is an arbitrary

[1] *B. Sl.* 1. 55 compare also *Āp. Sl.* V. 7.

number. Since p is also one side of the second rectangle, its other side $= \frac{1}{2}\left(\frac{p^2}{n} - n\right)$.

The two rectangles are juxtaposed and then right triangles are sliced off from either end of the resultant rectangle,

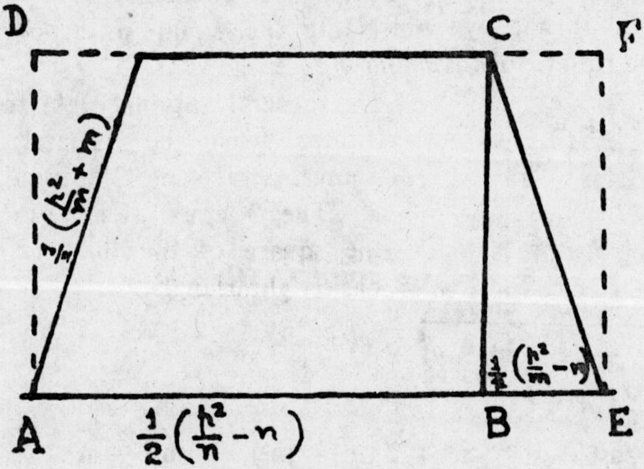

Fig. 15

Hence the base = sum of the koṭis $= \frac{1}{2}\left(\frac{p^2}{n} - n\right) + k$

and face = difference of the koṭis $= \frac{1}{2}\left(\frac{p^2}{n} - n\right) - k$

and the flanks are the diagonals of the first rectangle.

6.11.1.4. *To construct a rational trapezium with three sides equal*

The same device is used. But since the top also has now to be equal to the diagonal of the first rectangle, a rectangle with one side equal to the diagonal of the first rectangle and the other side equal to one side of the first rectangle is placed in the middle, and halves of the first rectangle are attached on either side of this. Besides, the general rational rectangle is derived from the most general right

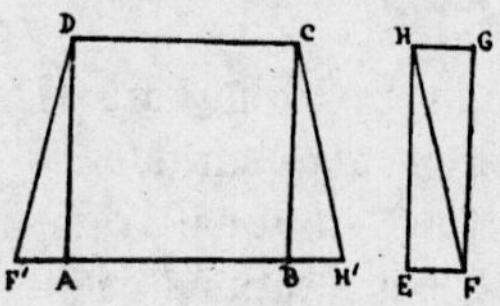

Fig. 16

triangle namely the one with sides m, n (i.e. with m and n as *iṣṭas*). Hence the rule

कर्णकृतिस्त्रिसमभुजास्त्रयः चतुर्थो विशोध्य कोटिकृतिम् ।
बाहुकृतेस्त्रिगुणायाः यद्यधिको भूर्मुखं हीनः ॥

(XII. 37)

(The three equal sides are the square of the diagonal and the fourth side is got by subtracting the square of the *koṭi* from thrice the square of the *bhuja* . . . . . .)

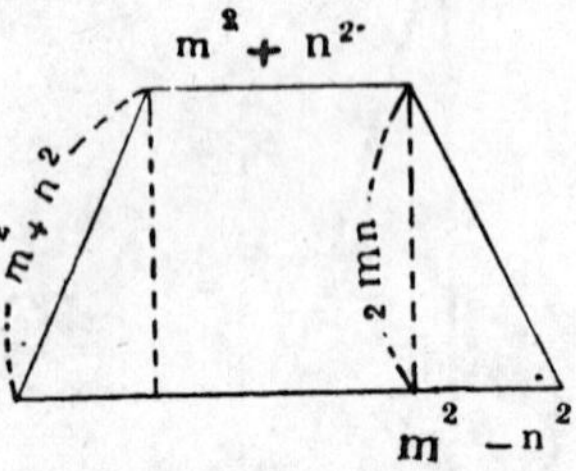

The general rational right triangle formed from the general right triangle is $m^2 - n^2$, $2\,m\,n$, $m^2 + n^2$

$\therefore$ The 3 equal sides $= m^2 + n^2$, the square of the diagonal of the basic right triangle.

Fig. 17

base $= m^2 + n^2 + 2\,(m^2 - n^2) = 3\,m^2 - n^2$.

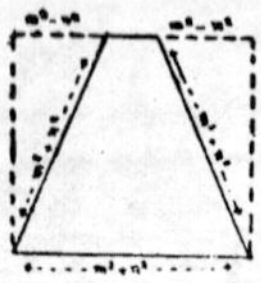

If this is less than $m^2 + n^2$, this will be the shorter parallel side and in this case the two right angled triangles are to be removed from either end of the central rectangle (Fig. 18).

Fig. 18

6.11.1.5. *To construct a rational quadrilateral*

The method is indicated in

जात्यद्वयकोटिभुजाः परकर्णगुणाः भुजाश्चतुर्विषमे ।

(XII. 38)

(The *koṭis* and *bhujas* of two rational right triangles multiplied by each other's hypotenuses are the four sides in a quadrilateral with unequal sides.)

How was the quadrilateral to be actually built up? Bhāskara II gives the same prescription for constructing a quadrilateral. Gaṅeśa, commenting on this, says that four rational right triangles are to be formed out of the two basic rational right triangles by multiplying the sides by the *bhuja* and *koṭi* of each other and the four are to be put together, so that their hypotenuses from the sides of the quadrilateral and the sides about the right angle combined two by two form the diagonals, i.e. if $m^2 - n^2$, $2mn$, $m^2 + n^2$ and $p^2 - q^2$, $2pq$, $p^2 + q^2$ are the two rational right triangles

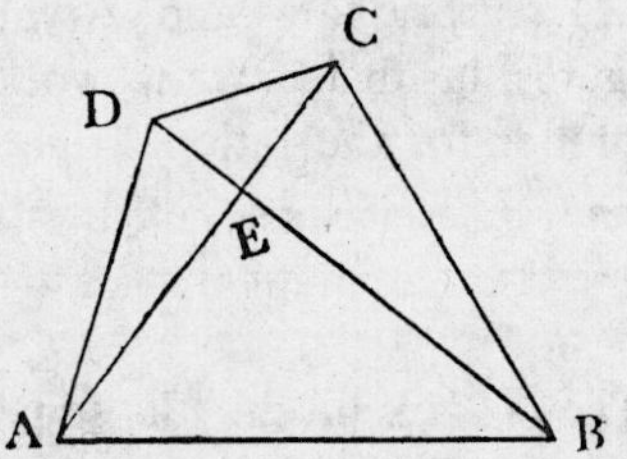

Fig. 19

the triangles out of which the quadrilateral is built up are

1) $(m^2 - n^2)(p^2 - q^2)$, $2mn(p^2 - q^2)$, $(p^2 - q^2)(m^2 + n^2)$
2) $(m^2 - n^2)2pq$, $4mnpq$, $2pq(m^2 + n^2)$
3) $(p^2 - q^2)(m^2 - n^2)$, $2pq(m^2 - n^2)$, $(p^2 + q^2)(m^2 - n^2)$
4) $(p^2 - q^2)2mn$, $4pqmn$, $(p^2 + q^2)2mn$.

But one wonders whether this is the procedure intended by Brahmagupta and Bhāskara, both of whom speak of multiplying the sides of two rational triangles by each other's hypotenuse and do not even hint at the need for four triangles. The anonymous commentary on the *Tantrasāra* of Nārāyaṇa (see 5.5.2) which contains much of the mathematics and astronomy of the later Āryabhaṭa School, quotes the above mentioned *Līlāvatī* verse and explains how the quadrilateral is to be built up. Two different rational right triangles (say 3, 4, 5 and 5, 12, 13) are selected and from them two other triangles are obtained by multiplying the sides by the other's hypotenuse (i.e. the new triangles will be 39, 52, 65, & 25, 60, 65). The two new triangles will have the same hypotenuse and can therefore be juxtaposed with their hypotenuses coinciding. Then one diagonal will be the common hypotenuse which will also be the diameter of the circumscribing circle. When the sides are interchanged other cyclic quadrilaterals are obtained. In the position of the sides in which no diagonal is diameter, we get the diagonals given by Bhāskara.

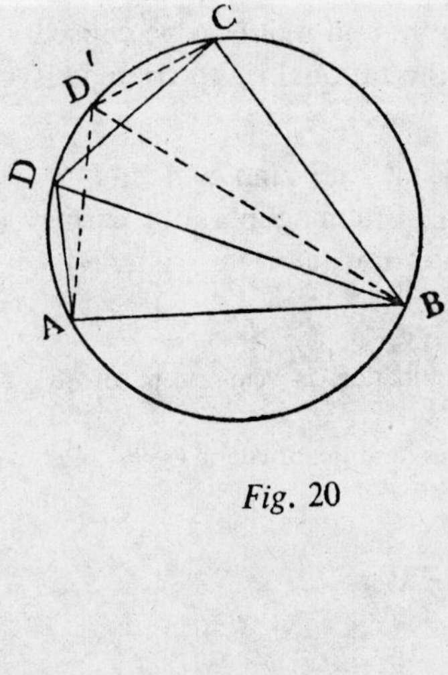

Fig. 20

6.11.2. Śrīdhara and Āryabhaṭa II do not treat of rational figures in their extant works. Śrīpati gives Brahmagupta's solution for the rational right triangle with the *bhuja* given.

इष्टा भुजा तत्कृतिरिष्टभक्तहीनार्द्धिता कोटिरसौ समेता ।
प्राग्भाजकेन श्रवस्सुधीभिर्जात्यो यतः क्षेत्रविधौ निरुक्तः ॥

(*Si. Se.* p. 87)

(The *bhuja* is given. Its square divided by an optional number and then diminished by the same number and halved is the *koṭi*. The same quotient combined with the above divisor is the hypotenuse. The learned have thus derived the *jātya* in geometry.)

The formation of the rational cyclic quadrilateral is also dealt with (*Si. Se.* p. 87).

6.11.3.1. Mahāvīra coming between Śrīdhara and Āryabhaṭa II accords a very full treatment to rational figures under a separate heading *Janyavyavahāra*. First of all he gives the general solution for the rational right triangle which is to be the basis for all that follows.

वर्गविशेषः कोटिः संवर्गो द्विगुणितो भवेद् बाहुः ।
वर्गसमासः[1] कर्णश्चायतचतुरश्रजन्यस्य ॥

(*G.S.S.* VII. $90\frac{1}{2}$)

(The difference of the squares is the upright side, twice the product is the horizontal side and the sum of the squares is the diagonal in a rectangle formed from *bījas*, elements.) As Dr. Datta points out,[2] Mahāvīra indicates, for the first time in the known history of Indian mathematics, the method of arriving at the results he gives. Any two integers m and n are to be chosen as the *bījas* or elements from which the rational right triangle is to be produced.

Then the solution for the triangle is $m^2 - n^2$, $2mn$ and $m^2 + n^2$. It is noteworthy that Mahāvīra, like Brahmagupta and earlier the *Śulbasūtras*, speak of the rectangle, not the right triangle.

[1]The use of *samāsa* here in the sense of combination is reminiscent of the *Śulbasūtra* practice.

[2]Mahāvīra's treatment of rational triangles and quadrilaterals—*Bull. Cal. Math. Soc.* 1930. p. 267.

The solution[1] for a rational right triangle with a side containing the right angle given, is the same as that of Brahmagupta.

Rational right triangle with the hypotenuse given is solved in

अथवा श्रुतीष्टकृत्योरन्तरपदमिष्टमपि च कोटि-भुजे ।

(*G.S.S.* VII. 97½)

(Or the root of the difference of the squares of the hypotenuse and an optional number and the optional number itself are the *bhuja* and the *koṭi*.) If c is the hypotenuse and m the chosen number, the sides are $\sqrt{c^2 - m^2}$, m and c. This is a rather unsatisfactory direct application of the Pythagorean theorem. Unless m is suitably selected, the triangle may not be possible at all, much less rational.

6.11.3.2. *Rational isosceles and scalene triangles*

The methods[2] closely follow Brahmagupta's. But Mahāvīra lets us know how the *bījas* for the two rational right triangles which are to be juxtaposed and therefore have to have one side about the right angle the same, are to be found.

6.11.3.3. *Rational isosceles trapezium*

The method[3] is essentially the same as Brahmagupta's. But where Brahmagupta gives the sides in terms of the sides of one rectangle and an optional quantity without explaining the method of formation, Mahāvīra makes it clear that two rectangles are to be used and tells us how to get *bījas* for the two right triangles, so that they may have a common side. The *bījas* for the second triangle are जन्यक्षेत्रभुजार्द्धहारफले i.e. the rational integral diviser and quotient of half the horizontal side of the first *janya*. That is, if m & n are the *bījas* of the first triangle, so that the *bhuja* is 2 mn, the *bījas* for the other will be

$$\frac{mn}{p} = q \text{ and } p.$$

[1] *G.S.S.* VII. 97½.
[2] *GS.S.* VII 108½ & 110½.
[3] *G.S.S.* VII 99½—100½.

6.11.3.4. *Rational trapezium with three equal sides*

भुजपदहतबीजान्तरहृतजन्यधनाप्तभागहाराभ्याम् ।
तद्भुजकोटिभ्यां च द्विसम इव त्रिसमचतुरश्रे ॥

(*G.S.S.*, VII. 101$\frac{1}{2}$)

(The quadrilateral with three sides equal is to be formed like the isosceles trapezium from two rectangles formed (1) with the quotient got by dividing the area of the given rational rectangle by the difference of its *bījas* as multiplied by the square root of its *bhuja*, and the divisor and (2) with the *bhuja* and *koṭi* of the first rectangle as *bījas*.)

The solution is the same as Brahmagupta's, whose wording tells us more clearly that the method is to have two rectangles such that the diagonal and one side of one are respectively equal to the two sides of the second.

6.11.3.5. *Rational cyclic quadrilateral*

The only difference between Mahāvīra's solution and Brahmagupta's is that Mahāvīra seems to multiply the sides of the two rational right triangles multiplied by each other's hypotenuse, again by the shorter diagonal. The purpose of this operation is not clear.[1] Nārāyaṇa Paṇḍita gives the same solution as Mahāvīra. Commenting on this, the editor of the *Gaṇitakaumudī* says that all the other elements of the quadrilateral, like the altitudes, the *ābādhās* and the circum-diameter, when computed in the quadrilateral as solved by Brahmagupta and Bhāskara, have the smaller hypotenuse as denominator, and so, to avoid fractions, Nārāyaṇa recommends the multiplication of all the elements including the sides by the smaller diagonal. This is plausible and the first to solve the quadrilateral with all elements integral was perhaps Mahāvīra.

6.11.4. Bhāskara's enunciation and elaboration of the theorem of the square of the hypotenuse is intended to make its useful-

[1]VII 103$\frac{1}{2}$ and 105 $\frac{1}{2}$-107$\frac{1}{2}$. Dr. Datta does not find any supefluous operation in Mahāvīra's prescription, but remarks that though Mahāvira follows Brahmagupta's method, his solution differs. He also credits Bhāskara with improving Brahmagupta's result, which improvement I fail to see.

ness in solving everyday problems and the problems of astronomy clear. At the cost of some verbosity he shows how from any two sides of a right triangle the third can be calculated. In his treatment of rational figures, which he confines to rational right triangles, he has a new solution for the sides of a right triangle with one side containing the right angle given.

इष्टो भुजोऽस्माद्द्विगुणेष्टनिघ्नादिष्टस्य कृत्यैकवियुक्तयाप्तम् ।
कोटिः पृथक् सेष्टगुणा भुजोना कर्णो भवेद्व्यस्त्रमिदं तु जात्यम् ॥

(*Līl.* 141)

(The base is given. This when multiplied by twice an optional number and divided by the square of that optional number diminished by one gives the upright (*koṭi*) side. That (*koṭi*) separately multiplied by the optional number and diminished by the *bhuja* is the hypotenuse. This is the rational right angled triangle (*jātya*). If a is the given side and m any optional number the *koṭi* is $\frac{2am}{m^2-1}$

and the *karṇa* is $\frac{2am^2}{m^2-1}-a$

Bhāskara's commentator, Sūraydāsa (16th century) explains how this solution is arrived at.[1] One solution of the rational right triangle is given by 2n, n^2-1 and n^2+1. Let a similar right triangle have base = a. Then its upright side

$$=\frac{a}{n^2-1}\times 2n=\frac{2an}{n^2-1}.$$

Again

n × the *koṭi* of the first triangle $= 2n^2 = (n^2-1)+(n^2+1) =$ its base + its hypotenuse ∴ n × the *koṭi* of the 2nd triangle = its base + its hypotenuse ∴ Its hypotenuse = n × *koṭi* — base

$$=n\times\frac{2an}{n^2-1}-a$$

By the same principle, if the hypotenuse c is given, the solution will be

[1]Vide H.C. Bannerji—Colebrooke's Translation of the *Līlāvati* with translation and notes, under 139.

$$\frac{2\,c\,n}{n^2+1},\ c-\frac{2\,cn^2}{n^2+1}\ \text{and}\ c$$

(*Līl.* 144)

This solution is a great improvement on Mahāvīra's solution. Bhāskara does not deal with other rational figures as such, though the solution of the rational cyclic quadrilateral is given in connection with the calculation of the diagonal of a cyclic quadrilateral.

6.11.5.1. At the very beginning of his section on *jātyakṣetras* Nārāyaṇa Paṇḍita makes the method to be followed very clear with the words.

भुजवर्गः श्रुतिकोट्योर्वर्गविशेषेण जायते तुल्यः ।
अन्तरमिष्टं कल्प्यं कोटिश्रवणौ ततो ज्ञेयौ ॥

(*G.K. Ks. Vya.*, 78)

(The square of the *bhuja* is equal to the difference of the squares of the hypotenuse and *koṭi*. The difference (between the hypotenuse and *koṭi*) should be assumed to be equal to an arbitrary number. Then the *koṭi* and the hypotenuse are to be calculated from these)[1]

$$\text{i.e. } a^2 = c^2 - b^2$$

$$\text{Let } c - b = m$$

$$\text{Then } c + b = \frac{a^2}{m}$$

$$\therefore c = \tfrac{1}{2}\left(\frac{a^2}{m} + m\right)$$

$$\text{and } b = \tfrac{1}{2}\left(\frac{a^2}{m} - m\right)$$

Nārāyaṇa himself explains the calculation in his attached notes. In verse 76 Bhāskara's solution for a rational right triangle with a given side viz. a, $\frac{2\,an}{n^2-1}, \frac{2\,an^2}{n^2-1} - a$ is given. Verses 80-81 give the solutions for the right triangle with the hypotenuse given. In verse 83 Nārāyaṇa gives a new garb, to the old solution 2n, n^2-1, n^2+1. His solution is m^2-n^2

[1]The method has been already explained conjecturally in connection with Brahmagupta's solution.

$\frac{(m-n)\{(m+n)^2-1\}}{2}, \frac{(m-n)\{(m+n)^2+1\}}{2}$, in which (m+n) replaces n and the whole solution is multiplied by (m—n) and divided by two. The mode of derivation is to assume that one side b in the triangle is equal to m^2-n^2. This solution is much more general than 2 n, n^2-1 and n^2+1.

6.11.5.2. The construction of rational isosceles trapezia in the *Gaṇita Kaumudī* follows the principles laid down by earlier mathematicians. But the author has a new type of trapezium, *karṇabhūsama* in which the diagonals are equal to the base. The rule for its construction comprehending many types of trapezia is being quoted in full.

श्रुतिबाह्वोः श्रतिकोट्योर्योगवियोगौ पृथक् पृथग्गुणितौ ।
भुजकोटिभ्यां करणीबीजे प्रथमाभिधे च भुजकोटी ॥
प्रथमभुजभवे ताभ्यां चतुरस्रं त्रिसमबाहुकं भवति ।
प्रथमजकोटिभवाभ्यां त्रिसमं वा कर्णभूसमं वापि ॥
बाहुजकोटिभवाभ्यां भूमिसमव्यासकं च चतुरश्रम् ।
द्विसमचतुरस्रविधिना भुजकर्णादीनि साध्यानि ॥

(*G.K.Ks., Vya.* 88-90)

(The sum and difference of (1) the diagonal and *bhuja* and (2) the diagonal and *koṭi* of a given rational rectangle is multiplied by the *bhuja* and *koṭi* separately. The square root of these products are two sets of *bījas*. The *bhuja* and *koṭi* themselves form another set of *bījas* called the *prathama* (first). The trapezium composed out of the two rectangles got from the *prathama* and *bhuja* sets of *bījas* will have three sides equal; the one from the rectangles with the *prathama* and *koṭi* sets of *bījas* will either have three sides equal or else have its base equal to its diagonals. The trapezium from the rectangles with the *bhuja* and *koṭi* sets of *bījas* will have its base equal to its circum diameter. The *bhujas*, diagonals etc. are to be obtained according to the instructions given for the isosceles trapezium.)

From a given rectangle with sides and diagonals a, b, c three sets of *bījas* are obtained.

1. *Bhuja* set. . . $\sqrt{(c+a)a}$ and $\sqrt{(c-a)a}$
2. *Koṭi* set . . . $\sqrt{(c+b)b}$ and $\sqrt{(c-b)b}$
3. *Prathama* set. . . a and b.

By forming rectangles with these *bījas* and combining them two by two by the usual method for building up trapezia, three kinds of trapezium are obtained

a) With three sides equal from (1) and (3)

The *jātya* obtained from (1) will have as sides and diagonal

$(c+a)a-(c-a)a=2a^2$

$2\sqrt{(c+a)a}.\sqrt{(c-a)a}=2\sqrt{a^2(c^2-a^2)}=2\sqrt{a^2b^2}=2ab$. and

$(c+a)a+(c-a)a=2ac$

The *jātya* obtained from (3) is a^2-b^2, $2ab$, a^2+b^2

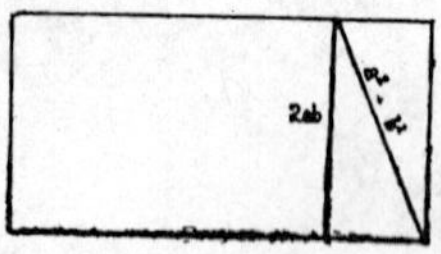

Fig. 21

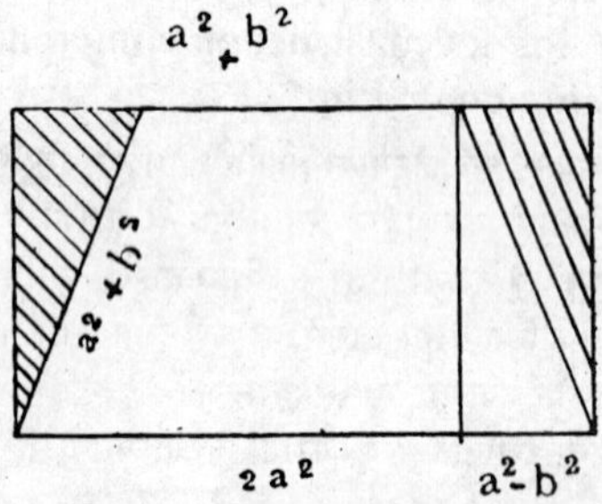

Fig. 22

Hence the face of the trapezium $=2a^2-(a^2-b^2)$

$=a^2+b^2=$ the flanks.

Therefore we get a trapezium with three sides equal.

(b) The *jātya* from the *koṭi* set of *bījas* will be $2b^2$, $2ab$, $2bc$

This combined with the *jātya* a^2-b^2, $2ab$, a^2+b^2 will give a trapezium with its base equal to its flanks.

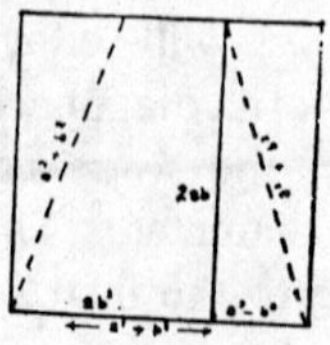

Fig. 23

If the diagonals of the trapezium are to be equal to the base, the diagonal of the smaller rectangle viz. a^2+b^2 is to be the diagonal of the trapezium and then the diagonals of the bigger rectangle which are equal to 2bc are to be the flanks. For this, triangles equal to half the bigger rectangle (i.e. ACD & EC'F)

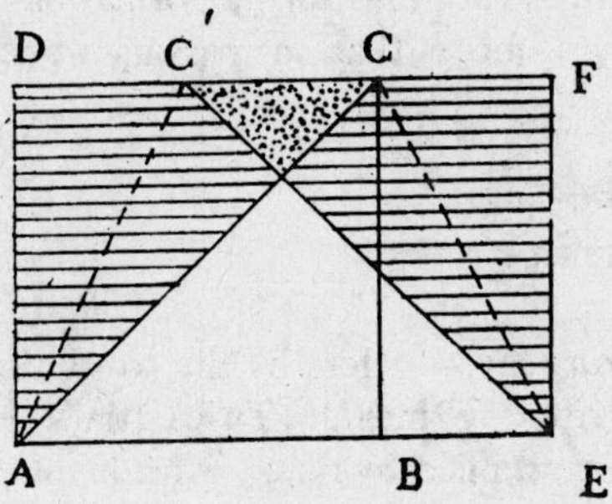

Fig. 24

are to be removed from either side of the combined rectangle.[1] Then naturally, the flanks being greater than the diagonals of the trapezium, they cross over each other and the trapezium takes the shape of two isosceles triangles joined at their apexes and the upper triangle is negative. Nārāyaṇa has provided for such a trapezium, most probably because in his section on the geometrical representation of series (*Średhīkṣetras*) such trapezia make their appearance. As he has explained in that context, the trapezium takes this shape (of the *kartari*-scissors as Nārāyaṇa puts it) because its face has a negative value.

(c) When the rectangles from the *bhuja* and *koṭi* sets of *bījas* are combined, the base will be $2a^2+2b^2$ and the circum-diameter

$$= \frac{\text{product of the diagonals of the } j\bar{a}tyas}{\text{common altitude}}$$

$$= \frac{2ac.2bc}{2ab} = 2c^2 = 2(a^2+b^2)$$

Hence the base = the circum-diameter.

6.11.6. The treatment of the formation of Brahmagupta's rational cyclic quadrilateral by the Āryabhaṭa School has been already dealt with. For the rest, the school does not devote much attention to rational figures. But the way in which the *Kriyā*-

[1]There is some ambiguity in Nārāyaṇa's own notes on this trapezium. He says (*G.K.*, Part II. p. 108).

अथ कर्णसमभूमिकानयनं जात्य प्रथमकोटिजम् । आभ्यां कर्तरी समम् । भूमिकम् कर्णौ १६६ १६६ लम्बौ । १२० १२० सन्धी ५० ५० पीठे ११६ ११६ व्यासः २१६७/१२ गणितम् ८०

Then two right triangles of sides 119, 120, 169 and 5, 42, 13 are shown. On the other hand the right triangles (or rectangles) he uses to illustrate the other types of trapezia are the three got from the triangle 3, 4, 5 by the method given in his rule i.e. 24, 32, 40; 18, 24, 30 and 7, 24, 25.

kramakarī interprets Bhāskara's formulas for the formation of rational right triangles deserves more than a passing notice. Commenting on

इष्टो भुजोऽस्माद्द्विगुणेष्टनिघ्नात् इष्टस्य कृत्यैकवियुक्तयाप्तम् ।
कोटिःपृथक् सेष्टगुणा भुजोना कर्णो भवेत्त्र्यस्रमिदं तु जात्यम् ॥

(*Līl.* 141)

the author of the *Kriyākramakarī* says—"Here by the word *iṣṭa* (chosen number), the *śara* (arrow) is meant. Twice the *koṭi* (perpendicular side) is to be multiplied by that. And if this is the *bhuja* (base) divided by the *śara*, twice the *koṭi* multiplied by the *śara* should be divided by the square of the *śara* or twice the *koṭi* by the simple *śara*. Here, primarily it is the sum of the *koṭi* and *karṇa* which is to be divided, but twice the *koṭi* is actually being divided and this is less than the sum of the *koṭi* and the *karṇa* by an amount equal to the *śara*. To decrease the divisor proportionately, one is being subtracted from it. Hence is it said 'divided by the square of the *iṣṭa* diminished by one'."

That is, the *K.K.* gets the sides of a rational triangle with one side given as follows. If a, b, are the sides about the right angle and c the hypotenuse,

$$b = \frac{b^2}{b} = \frac{c^2 - a^2}{b}$$

$$= \frac{c+a}{b/(c-a)}$$

$$= \frac{(c+a)\{(c+a)-(c-a)\}}{\frac{b}{c-a}\{(c+a)-(c-a)\}}$$

$$= \frac{(c+a)-(c-a)}{\frac{b}{c-a} - \frac{b}{c-a} \times \frac{c-a}{c+a}}$$

$$= \frac{2a}{m - m.\frac{c-a}{c+a}} \quad \left(\text{putting } \frac{b}{c-a} = m\right)$$

$$= \frac{2am}{m^2 - m^2.\frac{c-a}{c+a}}$$

$$= \frac{2a.m}{m^2 - \frac{b^2(c-a)}{(c-a)^2.(c+a)}} = \frac{2a.m}{m^2 - \frac{b^2}{c^2-a^2}}$$

$$= \frac{2a.m}{m^2 - \frac{b^2}{b^2}} = \frac{2a.m}{m^2-1}$$

And since $c+a = \frac{b^2}{c-a} = b.m,\ c = b.m - a.$

If we consider the triangle as made in a quadrant by the sine-chord (*bhuja*), cosine-chord (*koṭī*), and radius-hypotenuse (*karṇa*) of an arc, c—a=height of the arc (*śara*) and $\frac{bhuja}{śara} = \frac{b}{c-a}$ is the *iṣṭa*, the chosen number.

Or if, the *K. K.* goes on to say, the difference between the *karṇa* and the *koṭi* is accepted as the *iṣṭa* (m).

$$karṇa - koṭi = m$$

$$karṇa + koṭi = \frac{karṇa^2 - koṭi^2}{karṇa - koṭi}$$

$$= \frac{bhuja^2}{m}$$

Adding and halving, $karṇa = \frac{1}{2}\left(\frac{bhuja^2}{m} + m\right)$

$$= \frac{1}{2}\left(\frac{a^2}{m} + m\right)$$

Subtracting and halving, $koṭi = \frac{1}{2}\left(\frac{a^2}{m} - m\right)$

This is the second solution given by Bhāskara in the next verse (142). Thus both rules are formulated, according to the *K. K.*, in relation to the right triangle in a quadrant formed by the sine-chord and cosine-chord of an arc and its radius-hypotenuse.

The solution when the hypotenuse is given, contained in *Līl.* 144, is also derived by the author of the *K.K.* similarly with

$\frac{bhuja}{śara}$ or $\frac{b}{c-a}$ as the chosen number, m. (p. 497, 498)

Again commenting on

इष्टयोराहतिर्द्विघ्नी कोटिर्वर्गान्तरं भुजः ।
कृतियोगस्तयोरेव कर्णश्चाकरणीगतः ॥ *Lil.* 147)

The *Krīyākramakarī* says:

इदानीमिष्टप्रसंगेन वृत्तक्षेत्रगतानां भुजाकोटिकर्णानामानयने विशेषः प्रदर्श्यंते ।

(*K. K.* p. 507)

(Now in the context of *iṣṭas* the special features in arriving at the *bhuja*, *koṭi* and *karṇa* in a circle are shown.)
Then after explaining the literal meaning of the *sūtra* "Here since the two *bhujas* are equal, their *koṭis* also should be equal. Hence the reciprocal products of the *bhujas* and *koṭis* also should be equal. The *bhuja* and *koṭi* are the *iṣṭa* quantities. Hence it is said *iṣṭayorāhatir dvighnī* . . . multiplication of the chosen *bhujas* by the reciprocal *koṭis* is, anyway, to be done in addition and subtraction in a circle. That is why Mādhava has said

जीवे परस्पर निजेतरमौविकाभ्यां · ·

(quoted on p. 102)

If so, why is it not divided by the radius ? To get integral numbers. Where do we get such *bhujas* and *koṭis*? In the circle whose radius is the square of the radius of the circle. Hence it is well said *iṣṭayorāhutir dvighnī.* Here how will the *bhuja* be known? Since these *bhujas* and *koṭis* are to be mutually subtracted and the reciprocal *koṭis* are equal (to the *bhujas*), these are to be squared. Their difference will be the *bhuja* in that circle. That also will be rational because it is in the bigger circle. Hence it is said the base is the difference of the squares. Hence the *karṇa* will be the sum of the squares of the two *iṣṭas* chosen first. And all these will be rational."

Here the author of the *K. K.* gives a geometrical method for constructing the rational right triangle given by Bhāskara's solution. A right triangle is to be formed with base and perpendicular side equal to any two chosen numbers m and n. The circle of reference for this sine-cosine triangle (OAB) will be the one with the point O as centre and the hypotenuse OB as the radius. Draw another circle with the same centre and OD = OB^2 as radius. On the smaller circle mark off an arc XC= 2BX. Join OC and produce it to meet the outer circle in B′. Draw B′A′ perpendicular to OX.

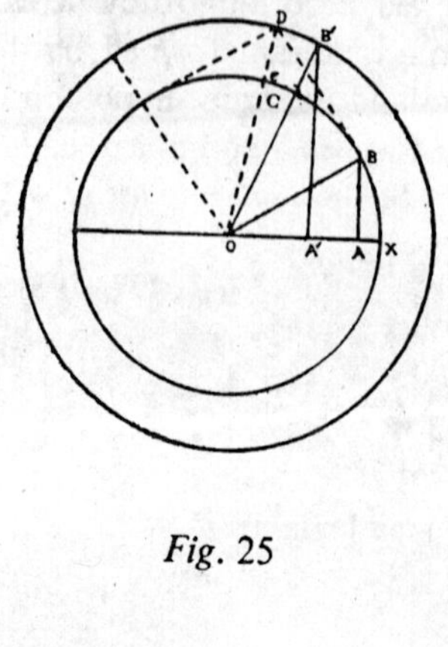

Fig. 25

Then OB′A′ is the rational right triangle corresponding to the solution 2mn, $m^2 - n^2$, $m^2 + n^2$.

Here also the rationale as explained by the *K.K.*, is trigonometrical. The sine and cosine chords for double the arc got by applying the summation formula for sine and cosine will be $\frac{2mn}{\sqrt{m^2+n^2}}$ and $\frac{m^2-n^2}{\sqrt{m^2+n^2}}$. To rationalise these and the hypotenuse $\sqrt{m^2+n^2}$, they are to be multiplied by $\sqrt{m^2+n^2}$ That is, if the radius is squared twice the same angle will have a rational sine-chord and cosine-chord and the hypotenuse or radius also will be rational. Hence the construction.

The school to which the *K. K.* belonged has a distinct bias towards geometry especially towards chord geometry and it succeeds well in linking these results belonging to the sphere of the Theory of numbers to chord geometry.

CHAPTER VII

THE CIRCLE

7.1. Indian geometry after the *Śulbasūtra* period grew up for the sake of the circle, the celestial circle. The *Śulbasūtras* attempt to square the circle and to circle the square (if the expression may be permitted). But the methods are but rough. The values of the ratio of the circumference to the diameter, implied in these methods are 3.004,3.0883, and 3.0885, slightly better than the ancient and very rough value 3.

The ancient Jainas did much more with the circle. They were aware that there is a fixed ratio between the diameter and the circumference and that the circumference multiplied by one-fourth the diameter is the area of a circle. The *Sūryaprajñapti* records the use of 3 as the value of π only to condemn it, while the approved value is $\sqrt{10}$. In the other early Jaina works like the *Jyotiṣkaraṇḍaka* and the *Jambūdvīpasamāsa* of Umāsvāti, π is invariably equal to $\sqrt{10}$, which value holds the field right up to the time of Bhāskara-II, though Āryabhaṭa I has a very good value for π. The probable origin of the value $\sqrt{10}$ has already been discussed. Amongst the Jainas too Vīrasena author of the *Dhavalā Ṭīkā* on the *Ṣaṭkhaṇḍāgama* gives a better value for π.

व्यासं षोडशगुणितं षोडशसहितं त्रिरूपरूपैर्भक्तम् ।
व्यासं त्रिगुणितं सूक्ष्मादपि तद्भवेत् सूक्ष्मम् ।

(*Ṣaṭkhaṇḍāgama* Vol. IV. p. 42)

(When the diameter multiplied by 16, combined with 16 and divided by 113 is again combined with thrice the diameter (the circumference) will be most exact.)

This will give the curious expression $\frac{16d + 16}{113} + 3\,d$. The constant term 16 in an expression for the circumference in terms of the diameter is illogical and if that 16 is removed we get the well-known and close approximation for π, $\frac{355}{113}$.

7.2. Āryabhaṭa's value of π, though well-known, bears quotation again.

चतुरधिकं शतमष्टगुणं द्वाषष्टिस्तथा सहस्राणाम् ।
अयुतद्वयविष्कम्भस्यासन्नो वृत्तपरिणाहः ॥

(*A. B. Gaṇitapāda* 10)

(The proximate value of the circumference for a diameter of 20000 is 62832.)

$$\text{i.e. } \pi = \frac{62832}{20000} = 3.1416$$

The value is quite a good one and yet Āryabhaṭa recognises that it is *āsanna*[1] only, a near approximation. Bhāskara's $\frac{3927}{1250}$ is Āryabhaṭa's value[2] with the common factor 16 removed from the numerator and denominator.

[1]Commenting on this word Nīlakaṇṭha Somayājin has a fine disquisition on the meaning of incommensurability. "कुतः पुनर्वास्तवीं संख्यामुत्सृज्यासन्नैवेहोक्ता । उच्यते । तस्या वक्तुमशक्यत्वात् । कुतः ? येन मानेन मीयमानो व्यासो निरवयवः स्यात् तेनैव मीयमानः परिधिः पुनः सावयव एव स्यात् । येन च मीयमानः परिधिनिरवयवस्तेनैव मीयमानो व्यासोऽपि सावयव एव, इत्येकनैव मानेन मीयमानयोरुभयोः क्वापि न निरवयवत्वं स्यात् । महान्तमध्वानं गत्वाप्यल्पावयवत्वं एवं लभ्यम् । निरवयवत्वं तु क्वापि न लभ्यमिति भावः ॥

"Why is this near value given here the real value being left out ? I will explain. Because the real value cannot be given. By the measure with which the diameter can be measured without a remainder the circumference measured by the same will certainly leave a remainder. Similarly the unit which measures the circumference without a remainder, will leave a remainder when used for measuring the diameter. Hence the two measured by the same unit will never be without a remainder. Though we carry it very far we can achieve smallness of the remainder only, but never remainderlessness. This is the idea."

[2]The *Kriyākramakarī* thinks that this is a gross value and says "केचित् पुनरत्रैवासन्नतरं परिधिमुद्दिश्य पाठान्तरं व्यधुः व्यासे शरेष्वग्निहते विभक्ते रामेन्दुरूपैः परिधिः सुसूक्ष्मः अयमेव पाठो युक्तिविदामभिमतः ॥

(Others have a variant reading that the diameter x $\frac{355}{113}$ is the very accurate circumference. This is the reading approved by people who know what mathematical reasoning is.)

7.3. The mode of arriving at this value was substantially the same as that for getting the value $\sqrt{10}$ but instead of stopping with the inscribed polygon of 12 sides, the number of sides was doubled till 384 was reached. Gaṇeśa suggests that the side of the 384-sides-polygon inscribed in a circle of diameter 100 was calculated by repeated application of the formula

$$S_{2n} = \sqrt{\frac{S_n^2}{4} + \left(r - \sqrt{\frac{4r^2 - S_n^2}{4}}\right)^2}$$

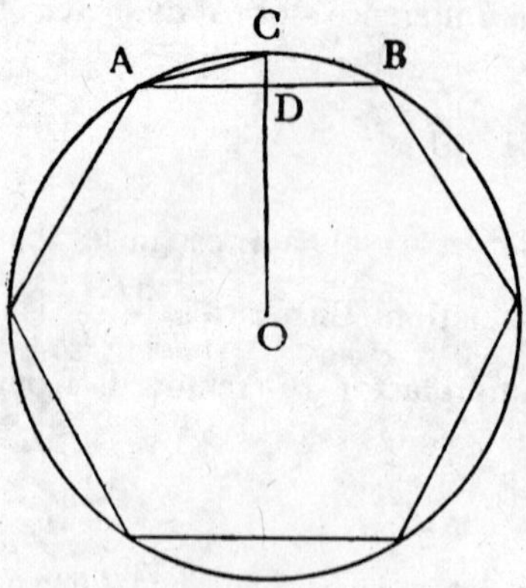

Fig. 1

where S_n and S_{2n} are the sides of the polygon of n sides and 2n sides respectively inscribed in the circle, the side of the inscribed hexagon being known to be equal to the radius. The *Yuktibhāṣā* employs for the same purpose the method of the escribed polygon, starting from the square and proceeding upto the polygon with a very large number of sides.[1]

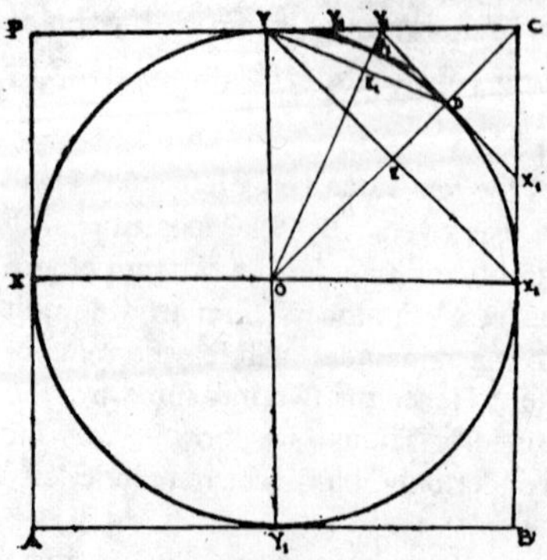

Fig. 2

Let ABCP be a square with side $=$ d, the diameter of the circle. Let the east-west line XOX′ and the north-south line YOY′ be drawn passing through the centre O of the circle and cutting the square into four smaller squares. Let OC and YX′, the diagonals of the square OX′CY be joined. Along OC set off OD $=$ YC and draw the perpendicular Y_1DX_1

Then from the similar triangles YEC and Y_1DC

$$\frac{Y_1C}{YC} = \frac{CD}{EC}$$

[1] *Y.B.* pp. 74-76

$$Y_1C = \frac{CD.YC}{EC}$$ ($EC = \frac{1}{2}\,OC$ and therefore can be calculated.)

And then $PC - 2Y_1C$ = the side of the octagon which will result if triangles similar to CY_1X_1 are cut off from the four corners of the square ABCD. Y_1Y is therefore a half side of the octagon.

$\therefore\ OY_1 = \sqrt{OY^2 + YY_1^2}$ can be calculated.

Let YE_1 be perpendicular on OY_1

Then $OE_1^2 - E_1Y_1^2 = OY^2 - YY_1^2$

i.e. $OY_1\,(OE_1 - E_1Y_1) = OY^2 - YY_1^2$

or $$OE_1 - E_1Y_1 = \frac{OY^2 - YY_1^2}{OY_1}$$

Hence E_1Y_1 can be calculated. Let OY_1 cut the circle in D_1. Y_2D_1 is drawn perpendicular to OY_1 touching YC in Y_2.

Then, as before, from the similar triangles, YE_1Y_1 and $Y_2D_1Y_1$, Y_2Y_1 can be calculated. Then the side of the polygon with 16 sides $= 2YY_1 - 2Y_2Y_1$.

In this way the number of sides of the polygon can be doubled and the corresponding side calculated till the number of vertices is very large and the polygon becomes a circle.

7.4. In these two methods, by increasing the number of sides of the polygon any desired degree of nearness can be achieved though the exact ratio can never be arrived at. A remarkably close approximation credited to Mādhava[4] by Nīlakaṇṭha Somayājin and others is in word-numerals.

बिबुध्र नेत्रगजाहिहुताशनत्रिगुणवेदभवारणबाहवः ।
नवनिखर्वमिते वृतिविस्तरे परिधिमानमिदं जगदुर्बुधाः ।।

The measure of the circumference in a circle of diameter 900,000,000,000 is 2,827,433,388,233.

(*K. K.* p. 668).

$$\text{i.e. } \pi = \frac{2,827,433,388,233}{900,000,000,000} = 3.14159265359$$

[1] *A.B. Gaṇitapāda* p. 42.

The *Krīyākramakarī* gives another as still closer.

वृत्तव्यासे हृते नागवेदवह्न्यब्धिखेन्दुभिः ।
तिथ्यश्विविबुधैर्भक्ते सुसूक्ष्मः परिधिर्भवेत् ॥

$$\text{i.e. } \pi = \frac{104348}{33215} = 3.1415926539211 \ldots$$

The *Karaṇapaddhati* (VI. 7) gives 31,415,926,536 as the circumference for a diameter of 10,000,000,000. The commentary published in the Madras edition of the *Karaṇapaddhati* (Madras Government Oriental Series, No. 98) shows how grosser approximations can be obtained from this by writing down the results of the continued division of the circumference and diameter and then working back with any desired number of results. The values of π thus obtained are

$$\frac{3}{1}, \frac{22}{7}, \frac{333}{106}, \frac{355}{113}, \frac{67783}{21576}, \frac{68138}{21689}, \frac{408473}{130021}, \text{ etc. (p. 176)}$$

7.5. The first of these methods is based on the theorem that the side of an inscribed hexagon is equal to the radius of the circle. This theorem must have been known in India quite early. Āryabhaṭa enunciates it.

परिधेः षड्भागज्या विष्कम्भार्धेन सा तुल्या ।

(*A. B. Gaṇitapāda*, 9)

(The chord of one-sixth of the circumference is equal to half the diameter).

The proof is given by the commentator Nīlakaṇṭha as also by the author of the *Yuktibhāṣā*.[1]

Let XOX′ be a diameter of the circle with O as centre and OX′ as radius. Mark a point A on the circum-

[1] *Y.B.* pp. 143-144.

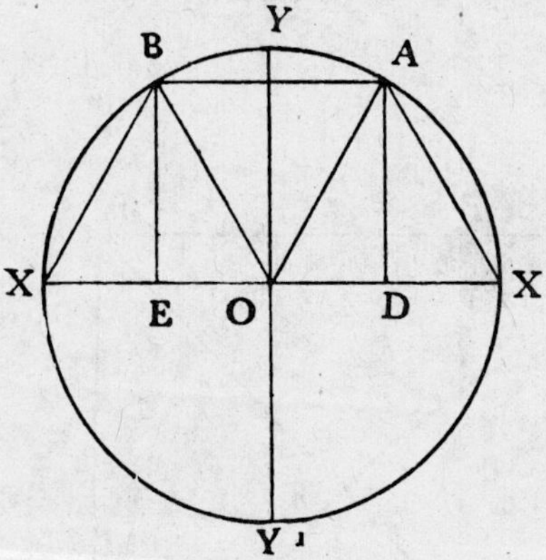

Fig. 3

ference such that chord X′A = OX′. OA, X′A are joined. Then OX′A is an equilateral triangle with side = OX′. A similar equilateral triangle is constructed on OX also with the vertex B on the circumference. Let AD and BE be drawn perpendicular to OX′ and OX from A and B. Then since OAX′ is an equilateral triangle, D is the middle point of OX′ Similarly E is the middle point of OX.

$$\therefore \quad ED = AB = \frac{1}{2}\,OX + \frac{1}{2}\,OX' = \frac{1}{2}XOX' = \frac{\text{diameter}}{2}$$

i.e. the semicircumference XBAX′ is divided into 3 parts whose chords are all equal to half the diameter. Hence the chord of the sixth part of the circumference is equal to half the diameter.

Āryabhaṭa II gives $\frac{22}{7}$ as the *sūkṣma* (accurate) value of π, whereas Bhāskara gives it as the *sthūla* (gross). To the author of the *Kriyākramakarī* it is *atisthūla*, very gross.[1]

7.6. *Series for* π

In addition to the methods of inscribing and escribing polygons, the Āryabhaṭa school uses a method of integration for getting values for π in the form of an infinite series. The series in its basic form is contained in

व्यासे वारिधिनिहते रूपहृते व्याससागराभिहते ।
त्रिशरादिविषमसंख्याभक्तं ऋणं स्वं पृथक्क्रमात् कुर्यात् ।

(*Tantrasamgraha* quotation in the *Y. B.* p. 99)[2]

(In the diameter multiplied by 4 and divided by one, decrease and increase should be made in turn of the diameter multiplied

[1]*K.K.* p. 668.

[2]A verse to the same effect occurs in the *Karaṇapaddhati* also (VI. 1)

by four and divided one by one by the odd numbers beginning with 3 and 5)

i.e. circumference $= 4d - \frac{4d}{3} + \frac{4d}{5} - \frac{4d}{7} + \ldots$

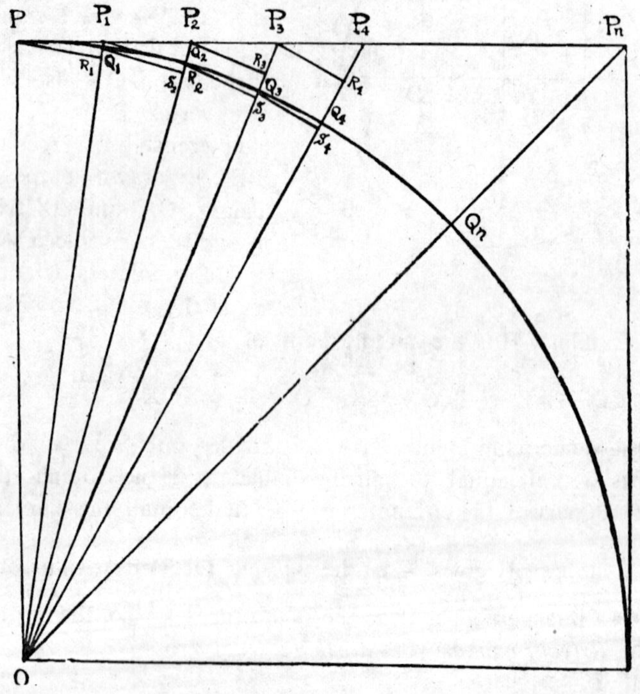

Fig. 4

The process of arriving at the series is a blending of geometrical reasoning and ingenious methods of summing up regular mathematical series.[1] The circle is inscribed in a square of side = the diameter of the circle, which will then touch the middle points of the sides of the square. A quarter of the circle with the circumscribing square is shown in the figure. The half-side PP_n of the square is divided into a number of very small equal parts PP_1, P_1P_2, P_2P_3 . . . . $P_{n-1}\,P_n$. The points P, P_1 . . . . P_n, are joined to the centre O, the joining lines OP_1, OP_2 . . . OP_n

[1] *Y.B.* pp. 85-99.

cutting the circumference in Q_1, Q_2, ... Q_n. These lines OP_1, OP_2 ... are termed *karṇas* by analogy with OP_n which is a diagonal. From P, P_1 ... P_{n-1} perpendiculars PR_1, P_1R_2 ... are dropped on the next *karṇas*. Similarly from Q_1, Q_2 ... also perpendiculars Q_1S_2, Q_2 S_3 ... are drawn.

Then from the similar triangles OPP_1 and OPR_1

$$\frac{PR_1}{PP_1} = \frac{OP}{OP_1}$$

$$\text{or } PR_1 = \frac{PP_1 \, . \, OP}{OP_1} = \frac{\Delta r.r}{OP_1}$$

(Where r is the radius of the circle and Δ r is the length of each small division on PP_n). Again from similar triangles P_1 R_2 P_2 and P O P_2

$$\frac{P_1R_2}{P_1P_2} = \frac{OP}{OP_2}$$

$$\text{i.e. } P_1R_2 = \frac{OP \, . \, P_1P_2}{OP_2} = \frac{\Delta r \, . \, r}{OP_2}$$

$$\text{Similarly } P_2R_3 = \frac{\Delta r \, . \, r}{OP_3}$$

$$P_3R_4 = \frac{\Delta r \, . \, r}{OP_4}$$

. . . .

. . . .

$$\therefore P_{n-1} R_n = \frac{\Delta r \, . \, r}{OP_n} \quad \ldots \quad \text{I}$$

Furthermore, triangles OP_1R_2 and OQ_1S_2 are similar.

$$\therefore \frac{Q_1S_2}{P_1R_2} = \frac{OQ_1}{OP_1}$$

$$\text{or } \quad Q_1S_2 = \frac{P_1R_2 \, . \, OQ_1}{OP_1}$$

$$= \frac{\Delta r \, . \, r}{OP_2} . \quad \frac{r}{OP_1} \text{ (Substituting from results I)}$$

$$= \frac{r^2 \, . \, \Delta r}{OP_1 \, . \, OP_2}$$

Similarly $\quad Q_2S_3 = \dfrac{r^2 . \Delta r}{OP_2 . OP_3}$

. . . .

. . . .

. . . .

$$Q_{n-1}S_n = \frac{r^2 . \Delta r}{OP_{n-1} . OP_n}$$

Now PR_1, Q_1S_2, Q_2S_3,........Q_{n-1} S_n are the half-chords (sine-chords) of the arcs PQ_1, Q_1Q_2.........Q_{n-1} Q_n. If the arcs are sufficiently small, i.e. the number of divisions on PP_n is sufficiently large, the arc can be taken to be equal to the half-chord. Hence arc PQ_n

$$= \tfrac{1}{8} \text{ of the circumference.}$$

$$= PR_1 + Q_1S_2 + Q_2S_3 + \ldots\ldots\ldots\ldots + Q_{n-1} S_n$$

$$= \frac{\Delta r.r}{OP_1} + \frac{\Delta r.r^2}{OP_1.OP_2} + \frac{\Delta r.r^2}{OP_2.OP_3} + \ldots\ldots + \frac{\Delta r.r^2}{OP_{n-1}.OP_n}$$

$$+ \frac{\Delta r.r^2}{OP.OP_1} + \frac{\Delta r.r^2}{OP_1.OP_2} + \ldots\ldots\ldots\ldots\ldots + \frac{\Delta r.r^2}{OP_{n-1}.OP_n}$$

Here since Δ r is assumed to be very small, OP_1 may be taken to be almost equal to OP_2.

i.e. $\quad OP_2 - OP_1 \to 0$

or $\quad (OP_2 - OP_1)^2 = OP_1^2 + OP_2^2 - 2OP_1 . OP_2 \to 0$

or $\quad OP_1^2 + OP_2^2 \to 2\, OP_1 . OP_2$

or $\quad \dfrac{1}{OP_1.OP_2} \to \dfrac{2}{OP_1^2 + OP_2^2}$

$$= \frac{2\,(OP_1^2 + OP_2^2)}{(OP_1^2 + OP_2^2)^2}$$

$$= \frac{2\,(OP_1^2 + OP_2^2)}{4\, OP_1^2 \times OP_2^2}$$

$$= \tfrac{1}{2}\left(\frac{1}{OP_1^2} + \frac{1}{OP_2^2}\right)$$

$$\therefore \frac{\text{Circum}^{ce}}{8} = \tfrac{1}{2}\left(\frac{\Delta r.r^2}{OP^2} + \frac{\Delta r.r^2}{OP_1^2}\right) + \tfrac{1}{2}\left(\frac{\Delta r.r^2}{OP_1^2} + \frac{\Delta r.r^2}{OP_2^2}\right) + \ldots$$

$$\ldots + \tfrac{1}{2}\left(\frac{\Delta r.r^2}{OP^2_{n-1}} + \frac{\Delta r.r^2}{OP_n^2}\right.$$

$$= \frac{1}{2}\left(\frac{\Delta r \,.\, r^2}{r^2} + \frac{\Delta r \,.r^2}{OP_1^2} + \frac{\Delta r \,.r^2}{OP_2^2} + \ldots\ldots + \frac{\Delta r \,.r^2}{OP^2_{n-1}}\right)$$

$$+ \frac{1}{2}\left(\frac{\Delta r.\, r^2}{OP_1^2} + \frac{\Delta r \,.r^2}{OP_2^2} + \;.\;.\;.\;.\;. + \frac{\Delta r \,.r^2}{OP_n^2}\right)$$

The difference between the two quantities within the brackets

$$= \frac{1}{2}\left(\frac{\Delta r \,.r^2}{r^2} - \frac{\Delta r \,.r^2}{OP_n^2}\right)$$

$$= \frac{1}{2}\left(\Delta r - \frac{\Delta r \,.r^2}{2r^2}\right) \text{ (since } OP_n^2 = OP^2 + PP_n^2\text{)}$$

$$= \frac{\Delta r}{2} - \frac{\Delta r}{4} = \frac{\Delta r}{4}, \text{ which is negligible}$$

$$\therefore \quad \frac{\text{Circum}^{ce}}{8} = \frac{\Delta r \,.r^2}{OP_1^2} + \frac{\Delta r.r^2}{OP_2^2} + \;.\;.\;.\;.\;. + \frac{\Delta r \,.r^2}{OP_n^2}$$

In these terms the denominators, (that is, the squares of the *karṇas*) OP_1^2, OP_2^2......are unknown. To eliminate these unknowns the denominator can be made the known square of the radius. The excess in the quotient resulting from the use of a smaller denominator is the quantity got by multiplying the quotient by the difference of the real denominator (*karṇa*²) and the assumed denominator (r^2) and dividing by the real denominator. To eliminate the unknown *karṇa*² from this subtrahend, it can be subjected to the same treatment. Thus by the repeated subjection of the subtrahends to the same treatment an infinite series involving the radius only is got from each of the terms.

$$\text{i. e. } \frac{\Delta r \,.r^2}{OP_1^2} = \frac{\Delta r \,.\, r^2}{r^2} - r^2.\frac{\Delta r\,(OP_1^2 - r^2)}{r^2 \,.\, OP_1^2}$$

$$= \Delta r - r^2 \,.\frac{\Delta r \,.\, (\Delta r)^2}{r^2 \,.\, OP_1^2}$$

$$= \Delta r - \frac{(\Delta r)^3}{r^4} \,.\, r^2 + r^2\,(\Delta r)^3\,\frac{(OP_1^2 - r^2)}{r^4 \,.\, OP_1^2}$$

$$= \Delta r - \frac{(\Delta r)^3}{r^2} + \frac{(\Delta r)^5}{r^4} - \frac{(\Delta r)^7}{r^6} + \ldots\ldots\ldots\ldots\ldots\ldots$$

$$\frac{\Delta r.r^2}{OP_2^2} = \frac{\Delta r \,.\, r^2}{r^2} - \frac{\Delta r\,(OP_2^2 - r^2)}{r^2 \,.\, OP_2^2}$$

$$= \Delta r - \frac{\Delta r\,(2\,\Delta r)^2}{OP_2^2}$$

$$= \triangle r - \frac{\triangle r(2\triangle r)^2}{r^2} + \frac{\triangle r(2\triangle r)^4}{r^4} \quad \ldots \quad \ldots$$

Similarly $\dfrac{\triangle r.r^2}{OP_3^2} = \triangle r - \dfrac{\triangle r(3\triangle r)^2}{r^2} + \dfrac{\triangle r(3\triangle r)^4}{r^4} - \ldots$

$$\frac{\triangle r.r^2}{OP_4^2} = \triangle r - \frac{\triangle r(4\triangle r)^2}{r^2} + \frac{\triangle r(4\triangle r)^4}{r^4} - \ldots$$

$$\ldots \quad \ldots \qquad \ldots \quad \ldots$$

$$\ldots \quad \ldots \qquad \ldots \quad \ldots$$

$$\frac{\triangle r.r^2}{OP_n^2} = \triangle r - \frac{\triangle r(n\triangle r)^2}{r^2} + \frac{\triangle r(n\triangle r)^4}{r^4} \ldots$$

$$\therefore \frac{\text{circumference}}{8} = \triangle r\left\{1 - \frac{(\triangle r)^2}{r^2} + \frac{(\triangle r)^4}{r^4} - \ldots \ldots \ldots\right\}$$

$$+ \triangle r\left\{1 - \frac{(2\triangle r)^2}{r^2} + \frac{(2\triangle r)^4}{r^4} - \ldots \ldots \ldots \ldots\right\}$$

$$+ \triangle r\left\{1 - \frac{(3\triangle r)^2}{r^2} + \ldots \ldots \ldots \ldots \ldots\right\}$$

$$\ldots \quad \ldots \quad \ldots \quad \vdots$$

$$\ldots \quad \ldots \quad \ldots \quad \ldots \quad \ldots \quad \ldots \quad \ldots$$

$$+ \triangle r\left\{1 - \frac{(n\triangle r)^2}{r^2} + \ldots \ldots \ldots \ldots \ldots\right\}$$

$$= n.\triangle r - \frac{\triangle r}{r^2}\left\{(\triangle r)^2 + (2\triangle r)^2 + (3\triangle r)^2 + \ldots \ldots \ldots\right\}$$

$$+ \frac{\triangle r}{r^4}\left\{(\triangle r)^4 + (2\triangle r)^4 + (3\triangle r)^4 + \ldots \ldots \ldots\right\}$$

$$- \frac{\triangle r}{r^6}\left\{(\triangle r)^6 + (2\triangle r)^6 + (3\triangle r)^6 + \ldots \ldots \ldots \ldots\right\}$$

$$+ \ldots \ldots \ldots \ldots \ldots \ldots \ldots \ldots$$

$$- \ldots \ldots \ldots \ldots \ldots \ldots \ldots \ldots$$

$$+ \ldots \ldots \ldots \ldots \ldots \ldots \ldots \ldots$$

$$- \ldots \ldots \ldots \ldots \ldots \ldots \ldots \ldots$$

To proceed further, one has to know the sums of the quantities within the brackets, i.e. the sums of the squares, 4th powers, 6th powers (*vargasaṃkalita, samacaturghāta-saṃkalita, samaṣaḍghātasaṃkalita*) etc. of the parts of a whole, which increase gradually by an infinitesimally small part. Therefore, as a necessary digression, the author shows how such parts raised to any power can be summed up. First, the series in the first degree is taken up

i.e. the series $\triangle r + 2\triangle r + \dots \quad \dots \quad \dots + n.\triangle r$

If all these parts were equal to r, their sum will be $r + r + \dots \dots$ If we assume $\triangle r = 1$ unit there will be r terms in the series i.e.

$n = r. \therefore$ The sum of $r + r + \dots \quad \dots \quad \dots \quad = r.r = r^2$

Writing the original series inversely underneath this and subtracting we get

$o + \triangle r + 2\triangle r + \dots \dots \dots + (r-1)\cdot\triangle r = r^2 -$ original series i.e. original series $- r = r^2 -$ original series

$$\therefore \text{ original series} = \frac{r^2+r}{2} = \frac{r(r+1)}{2}$$

If $\triangle r = 1$ unit is chosen sufficiently small, $r+1$ will not be appreciably different from r

$$\text{Hence } \triangle r + 2\triangle r + \dots \dots + r = \frac{r^2}{2}$$

Similarly it can be shown that

$$(\triangle r)^2 + (2\triangle r)^2 + (3\triangle r)^2 + \dots = \frac{r^3}{3}$$

$$(\triangle r)^3 + (2\triangle r)^3 + (3\triangle r)^3 + \dots \dots \dots \dots = \frac{r^4}{4}$$

and so on

$$\text{Hence } \frac{\text{circumference}}{8} = n.\triangle r - \frac{\triangle r}{r^2}\cdot\frac{r^3}{3} + \frac{\triangle r}{r^4}\cdot\frac{r^5}{5} - \frac{\triangle r}{r^6}\cdot\frac{r^7}{7} + \dots \dots \dots$$

$$= r - \frac{r}{3} + \frac{r}{5} - \frac{r}{7} + \dots \dots$$

$$\therefore \text{ circumference} = 8r - \frac{8r}{3} + \frac{8r}{5} - \frac{8r}{7} + \dots\dots$$

$$= 4d - \frac{4d}{3} + \frac{4d}{5} - \frac{4d}{7} + \dots\dots$$

The Āryabhaṭa school had worked out a method of applying a correction to the value got by taking any number of terms and with its help found the circumference in the *kaṭapayādi* system to be चण्डांशुचन्द्राधमकुंमिपाल! for a radius of श्रानूननून्नाननुन्ननित्यम्

$$\text{i.e. } \pi = \frac{31{,}415{,}926{,}536}{10{,}000{,}000{,}000}$$

From this they could get many other approximations to the value of π (see 7.5 above)

This series for $\frac{1}{8}$th of the circumference can be manipulated to yield several other series. Thus by calculating the length of $\frac{1}{12}$th of the circumference instead of $\frac{1}{8}$th we get the series[1]

$$C = \sqrt{12\,d^2} - \frac{\sqrt{12d^2}}{3.3} + \frac{\sqrt{12d^2}}{3^2.5} - \ldots{}^{2}$$

By grouping the terms differently new series have been derived out of $C = 4d - \frac{4d}{3} + \frac{4d}{5} - \ldots\ldots$

1. $C = 4d - 4d\left(\frac{1}{3} - \frac{1}{5}\right) - 4d\left(\frac{1}{7} - \frac{1}{9}\right) \ldots\ldots$

2. $C = 8d\left(\frac{1}{2^2-1} + \frac{1}{6^2-1} + \frac{1}{10^2-1} + \ldots\right)$

3. $C = 4d - 8d\left(\frac{1}{4^2-1} + \frac{1}{8^2-1} + \frac{1}{12^2-1} + \ldots\right)$

D.E. Smith in a foot-note on page 309 of his *History of Mathematics* Vol. II notices C. M Whish's article on the Hindu quadrature of the circle, but still does not mention these infinite series for π discovered in India latest by the 15th century, while he mentions (pp. 311-312) those discovered by John Wallis (1655 A.D.), Leibniz (1673). Abraham Sharp (c. 1717), John Machin (c. 1706) and Matsunga Ryohitsu (1739). The Yenri or Circle Principle of Japanese mathematics made its appearance in the 17th century. It will be interesting to investigate whether

[1] *Y.B.* pp. 117-119.

[2] This and the following series have been collected together along with the original Sanskrit verses as an appendix to *On the Hindu Quadratue of the circle*. by C.T. Rājagopal and Mukunda Mārar.

the Yenri had any connection with the circle mathematics of India of the 14th and 15th centuries. Probably because the series for π were discovered in the remote south-west corner of India, these are not found in the Arab, Persian or other foreign works of the medieval period.

7.7. *Area of a circle*

Once the circumference of a circle was determined, finding the area was easy. For from very early times it was known that the area of a circle = circumference . $\frac{d}{4}$. The early Jaina works like Umāsvāti's (c. 150 B.C.) *Tattvārthādhigamabhāṣya* are familiar with it, though they give no indication as to how the result was arrived at. Āryabhaṭa's statement of the result seems to be meant to give his readers a clue to the method.

समपरिणाहस्यार्धं विष्कम्भार्धहतमेव वृत्तफलम्

(*A. B. Gaṇitapāda*)

(Half the circumference multiplied by half the diameter is the area of a circle.)

The details are given by Nīlakaṇṭha. A circle can be cut up into a large number of *Śūcyākārakṣetras* (tapering figures) by means of lines drawn from the centre to the circumference. If

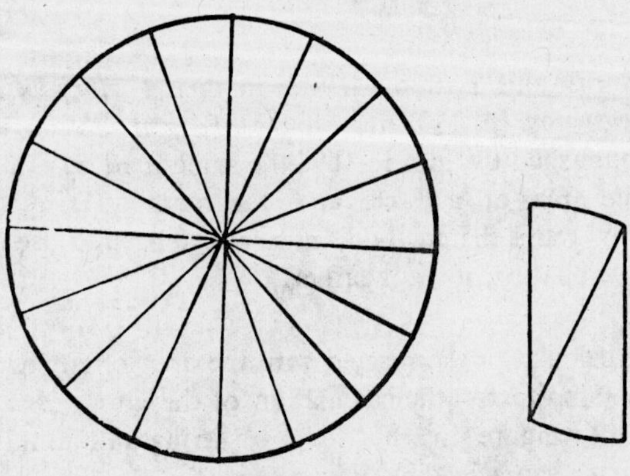

Fig. 5

the number of such *sūcis* is made sufficiently large, the base of these triangles will be straight lines. When two of these thin *sūcīs* are combined inverted, a rectangle with one side equal to the radius and the other side equal to the base of the *sūcīs* results. The whole circle can thus be re-arranged as thin rectangles. These thin rectangles attached one to another with the sides equal to the radius coinciding, will produce a rectangle with length = half the circumference and breadth = the radius.

Hence the area of the circle $= \frac{1}{2}$ circumference. $\frac{\text{diameter}}{2}$

The same proof with a slight modification is given by Gaṇeśa and the *Y.B.* The *sūcīs* are not to be separated out and put together again. Instead, the circle is cut into two semicircles first and the same number of radial lines are cut in the two halves reaching upto the circumference. Then the two pieces are straightened out and joined inverted to yield a rectangle with length $= \frac{1}{2}$ the circumference and breadth = the radius.

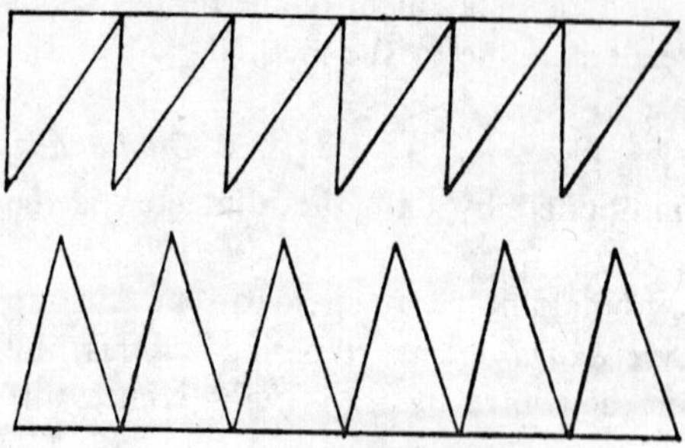

Fig. 6

D. E. Smith and Yoshio Mikami in their *History of Japanese Mathematics* say (p. 130) that in the *Tengan Shinan* of Sato Moshun published by him in 1698 the same method is used for finding the area of a circle and the method is distinctively western. In India the method must have been in use from the time of Āryabhaṭa I, if not earlier.

7.8. One calculation connected with the area of circles is peculiar to the Jaina texts—the calculation of the area of a *valayākārakṣetra* (a figure in the form of a ring), an annulus. The formula first occurs in the *Tiloyappaṇṇatti*.

दुगुणिच्चिय सूचीए इच्छियवलयाण दुगुणवासाणि।
सोधिअ अवसेसकिदिं वासद्धकदीहि गुणिदूणं॥

गुणिदूणं दसेहि तदो मूलेणंकं हवेदि जं लद्धं ।
इच्छियवलयायारे खेत्ते तं जाण सुहुमफलम् ॥

(IV 2521-2522)[1]

(Subtract twice the breadth of the desired annulus from twice the outer diameter and multiply the square of the remainder by the square of half the breadth and by 10. Know the square root of the product as the accurate area of the desired annulus). That is, if d is the outer diameter and t the breadth of the ring, the area

$$= \sqrt{10\left(\frac{t}{2}\right)^2(2d-2t)^2} = \frac{\pi t}{2}(2d-2t)$$

For area of annulus = area of outer circle — area of inner circle

$$= \frac{\pi d^2}{4} - \frac{\pi(d-2t)^2}{4}$$

$$= \pi.\ \frac{(d^2-d^2+4dt-4t^2)}{4}$$

$$= \frac{\pi t}{2}(2d-2t)$$

The same formula is given in the *Trilokasāra* with an insignificant variation (V. 315)

Area of annulus $= \frac{\pi}{2}(d_1+d_2)t$, where d_1 and d_2 are the outer and inner diameters and t is the breadth of the annulus.

$$(\text{area} = \frac{\pi}{4}(d_1^2-d_2^2) = \frac{\pi}{4}(d_1+d_2)(d_1-d_2) = \frac{\pi}{4}(d_1+d_2)2t$$

$$= \frac{\pi}{2}(d_1+d_2)t.)$$

Mahāvīra gives formulae for the area of an out-lying and in-lying annulus (*G.S.S.* VII. 28) as $A=(d\pm t)\,\pi.t$ (where d is the inner or outer diameter and t the width of the annulus). The

[1]The word *vyāsa* has no fixed meaning in Jaina literature. Here it means the breadth of the annulus. The Sanskrit rendering of the verses is

द्विगुणीकृतसूच्या: इष्टवलयानां द्विगुणव्यासान् ।
शोधयित्वा अवशेषकृतिं व्यासार्द्धकृत्या गुणयित्वा ॥
गुणयित्वा दशभि: ततो मूलेनांक: भवति यो लब्ध: ।
इष्टवलयाकारक्षेत्रे तं जानीहि सूक्ष्मफलम् ॥

gross area of a *nemi* is given as नेमेर्भुजयुत्यर्धं व्यासगुणम् (The area of a *nemi* is half the sum of the *bhujas* multiplied by the breadth). (*G.S.S.* VII.7). Though Prof. Rangacharya interprets *nemi* as an annulus, the word perhaps means *rathāṅgaśakala*" (bit of a wheel or annulus), which is the definition given to it by Nārāyaṇa Paṇḍita (*G.K.Ks*, *Vya.* 14). The exact area of a *nemi* is given as "The sum of the back side and the inner side divided by 6 and multiplied by the breadth and by the root of 10 is the area of the *nemi*' (*G.S.S.* VII. $30\frac{1}{2}$) i.e. area of $nemi = \frac{a_1+a_2}{6} t. \sqrt{10}$ where a_1 and a_2 are the inner and outer lengths of the *nemi*. The only explanation for this formula is that Mahāvīra considers the formula given in verse 7, gross because the gross value of π namely 3 is involved in it. Hence he divides the whole by 3 and multiplies by $\sqrt{10}$, his exact value of π. Nārāyaṇa Paṇḍita gives the expression for the area of an annulus as $(dt+t^2)\pi$ (*G.K.Ks. Vya.* 14). No separate expression is given for the area of a *nemi*.

Mahāvīra attempts to give expressions for the circumference and area of an ellipse also. His classification of regular closed curves into *samavṛtta* and *āyatavṛtta* (equal and elongated circles) is perhaps traceable to early Jaina literature. where we come across *samacakravāla* and *viṣamacakravāla*. The rules for calculating the circumference and area of an ellipse are : The *āyāma* (longer diameter), combined with half the *vyāsa* (shorter diameter) and doubled gives the circumference. One fourth the shorter diameter multiplied by the circumference gives the area." (*G.S.S.* VII. 21) i.e. if a and b are the major and minor axes of the ellipse, $\text{circumference} = 2(a + \frac{b}{2})$

and $\text{area} = \text{circumference}. \frac{b}{4} = \frac{1}{4} b.2(a + \frac{b}{2}) = \frac{1}{2} b(a + \frac{b}{2})$.

The formulae seem to be the result of an arbitrary extension of the ones for the circle to the ellipse. The rough value for the circumference of a circle is 3 d and of the 3 diameters, two are allotted to the major axis and one to the minor arbitrarily. The values given as exact for the circumference and area of the ellipse in VII. 63, viz. $\sqrt{4a^2+6b^2}$ and $\frac{b}{4}\sqrt{4a+6b^2}$ seem similarly to have been obtained by taking $\pi = \sqrt{10}$ and distributing 10 between a^2 and b^2. It seems unlikely that the properties of an ellipse were really investigated into.

7.9. Mahāvīra and Nārāyaṇa Paṇḍita give accurate formulae for the area of the space enclosed by three or more equal, mutually touching circles.

Fig 7

If the diameter of the circles is d, the area enclosed by 4 circles = area of a square with side equal to the diameter minus the area of one circle[1] $= d^2 - \frac{\pi d^2}{4}$.

In the case of three circles,[2] the enclosed area = area of an equilateral triangle with side equal to the diameter minus half the area of one circle. Here since the angles of the three sectors are 60° each, each of the sectors $= \frac{1}{8}$ the circle. ∴ 3 sectors together $= \frac{1}{2}$ the circle. Hence the rule.

Fig. 8

A general formula for the space enclosed by any number of equal circles is also attempted.

[1]*G.S.S.* VII $82\frac{1}{2}$.

[2]Ibid. $84\frac{1}{2}$.

रज्ज्वर्धकृतित्र्यंशो बाहुविभक्तो निरेकबाहुगुणः ।
सर्वेषामश्रवतां फलं हि बिंबान्तरे चतुर्थांशः ॥

(*G.S.S.* VII. 39)

(One third the square of the semi-perimeter divided by the number of sides and multiplied by the number of sides diminished by one gives the area of all regular figures with corners (or rectilinear sides) and one-fourth of this is the area enclosed between the circles.)

That is, if a polygon has n sides each equal to a

$$\text{its area} = \frac{\left(\frac{n\,a}{2}\right)^2}{3} \cdot \frac{(n-1)}{n}$$

$$\text{or } \frac{s^2}{3} \cdot \frac{n-1}{n}, \text{ where s is the semiperimeter.}$$

This is true of triangles, squares and the circle considered as a polygon with infinite number of sides, but only if the approximate values are taken. At any rate, the formula is a clever one.

The area enclosed by the circles at the vertices of a regular polygon

$$= \frac{1}{4} \cdot \frac{s^2}{3} \cdot \frac{n-1}{n}$$

This also works for 3 and 4 circles but not for higher numbers.

Nārāyaṇa has two similar formulas

रश्म्यूनरश्मिकृतिहतभुजकृतिरिनहृत् फलं त्रिकोणादौ ।

(*G.K.Ks.Vya.* 15)

(The square of a side multiplied by the square of the number of sides as diminished by the number of sides and divided by 12, gives the area in an equilateral triangle etc.)

i.e. the area of an equilateral polygon

$$= \frac{(n^2-n)a^2}{12}$$

This is really old wine in a new bottle.

For Mahāvīra's formula $A = \frac{s^2}{3} \cdot \frac{n-1}{n}$

$$= \frac{(na)^2}{3.4} \frac{(n-1)}{n}$$

$$= \frac{n^2a^2\,(n-1)}{12.n}$$

$$= \frac{(n^2-n)a^2}{12}$$

For the area enclosed by the circles at the vertices Nārāyaṇa has

व्याससमासार्धकृतिर्निरेकवृत्ताहता हृता वृत्तैः ।
नवगुणितैर्वृत्तान्तरफलमथवा रश्मिजं त्रिहृतम् ॥

(*G.K.Ks.Vya.* 16)

(The square of half the sum of the diametres multiplied by the number of circles diminished by one and divided by 9 times the number of circles, is the area enclosed by the circles; Or, it is the area got with the above *raśmi* (number of sides) formula divided by three)

i.e. the area enclosed $= \left(\frac{nd}{2}\right)^2 \cdot \frac{n-1}{9n}$ where d is the diameter and n the number of circles.

$$\text{or} = \frac{d^2\,(n^2-n)}{12.3}$$

The two formulae are almost the same. Only, where Mahāvīra has $\frac{1}{4}$ the area of the polygon, Nārāyana has $\frac{1}{3}$. The change is not for the better. The result is now strictly correct for no value of n. For, the formula gives too high a value for 3 and 4 circles and too low a value for numbers above 4.

7.10. The segment and the chord

The expressions for the height, the chord and the diameter of a segment in terms of the others occur in the *Tattvārthādhigama-bhāṣya* and other early Jaina works.

(1) $h = \frac{1}{2}\left(d - \sqrt{d^2 - c^2}\right)$ where h is the height, d the diameter and c the length of the chord.

For $h = CD = OC - OD$ (fig. 9)

$$= OC - \sqrt{OA^2 - AD^2}$$

$$= \frac{d}{2} - \sqrt{\frac{d^2-c^2}{4}}$$

(2) $c = \sqrt{4h(d-h)}$.

This is obviously obtained by a consideration of the similar triangles ADC′ and CDA, from which

$$\frac{CD}{AD}=\frac{AD}{C'D}$$

or $$\frac{h}{c/2}=\frac{c/2}{d-h}$$

or $$\frac{c^2}{4}=h(d-h)$$

or $$c=\sqrt{4h\,(d-h)}$$

(3) $$d=\frac{c^2/4+h^2}{h}$$

This follows from result (2)

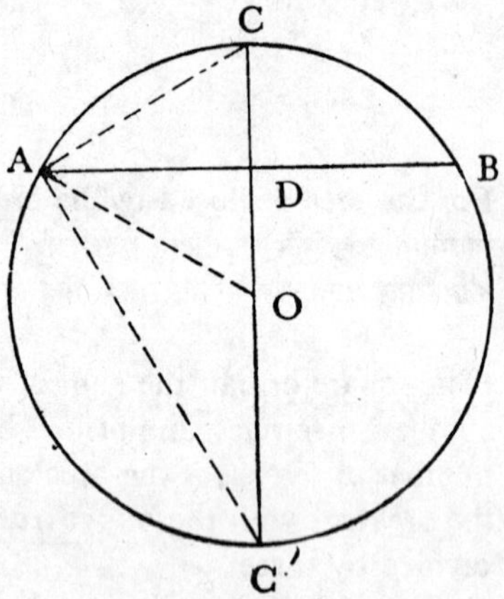

Fig. 9

The arc, the chord and the height of a segment are sought to be connected in another sequence of formulae

$$a=\sqrt{6h^2+c^2} \text{ (a is the length of the arc.)}$$

$$h=\sqrt{\frac{a^2-c^2}{6}}$$

$$c=\sqrt{a^2-6h^2}.$$

These expressions occurring in the *Tattvārthādhigamabhāṣya* etc. are very old and were known in ancient China. Even the *Upāṅgas*, a section of the oldest part of Jaina canonical literature are likely to have been aware of these or similar formulae. The *Jambūdvīpaprajñapti* gives the lengths of arcs to small fractions with the addition किंचिद्विसेससहिय्रं, with something still remaining (S.11 & 16). *Jīvā* (chord) also is mentioned. These formulae are probably the result of rough generalisations from the expression for the semi-circumference in terms of the diameter, the diameter lengths in the expression being arbitrarily distributed between the height and chord of the semi-circle. For

$$\text{semi-perimeter} \quad \sqrt{\frac{10\,d^2}{4}}$$

$$= \sqrt{\frac{6d^2 + 4d^2}{4}}$$

$$= \sqrt{6h^2 + c^2} \text{ since } h = \frac{d}{2} \text{ and } c = d$$

in a semi-circle. This was perhaps accepted as true for all segments.

The other expressions are but inversions of this. The Jainas have many other expressions too for the segment, which seem to have been similarly derived by arbitrary generalisation.

1. The area of a segment $= \sqrt{10}.\ \frac{c\ h}{4} = \frac{\pi\ c.\ h}{4}$ occurs in the *Bṛhatkṣetrasamāsa* of Jinabhadragaṇi (529 — 589 C)[1]

2. Length of arc $a = \sqrt{2 \left\{ (d + h)^2 - d^2 \right\}}$, occurring in the *Tiloyapaṇṇatti* (IV. 181) is of the same type.[2]

3. $a = \sqrt{4h\left(d + \frac{h}{2}\right)}$

area of segment $= \frac{1}{2}$ (c+h). h and their derivatives, viz.

$$d = \frac{1}{2}\left(\frac{a^2}{2h} - h\right)$$

$$h = \sqrt{d^2 - \frac{a^2}{2}} - d$$ (got by solving the

quadratic equation $a = \sqrt{4h\left(d + \frac{h}{2}\right)}$

The last three occur in the *Trilokasāra* of Nemicandra (10th century).

Āryabhaṭa I's statement of the theorem about the chord of a circle is characteristically compact.

[1]B.B. Datta — Mathematics of Nemicandra, *Jaina Antiquary. Bhāga* 2. *kiraṇa* 2. pp. 34-38. Also IV. 2374, *Tiloyapaṇṇatti.*

[2]For, semi-circumference $= \frac{\pi d}{2} = \frac{\sqrt{10}.d}{2}$

$\therefore$ semi-circumference^2

$$= \frac{10\ d^2}{4} = 2\left(\frac{9d^2}{4} - d^2\right) = 2\left\{\left(d + \frac{d}{2}\right)^2 - d^2\right\}$$

$$= 2\left\{(d + h)^2 - d^2\right\}$$

वृत्त शरसंवर्गो ऽर्धज्यावर्गः स खलु धनुषोः

(*A. B. Gaṇitapāda* 17)

(In a circle, the product of the arrows is the square of half the chord of the arcs.)

Brahmagupta gives this and the derived formulas (*Br. Sp. Si.* XII, 41 & 42). Neither of these mathematicians deals with the arc or the segment in the section on *Kṣetragaṇita*. But in the *Spaṣṭādhikāra*, the arc is calculated from the *ardhajyā* (half-chord) with the help of the sine-table. Śrīdhara's expression for the area of a segment is

जीवाशरैक्यदलहतशरस्य वर्गं दशाहतं नवभिः ।
विभजेदवाप्तमूलं प्रजायते कार्मुकस्य फलम् ॥ (*T. S.* 47)

(The square of the arrow as multiplied by half the sum of the chord and the arrow should be multiplied by 10 and divided by 9. The square root of the quotient gives the area of the segment.)[1]

$$\text{i.e. area of segment} = \sqrt{\left\{\frac{h\,(c+h)}{2}\right\}^2 \cdot \frac{10}{9}}$$

$$= \frac{\pi}{3}\sqrt{\left\{\frac{h\,(c+h)}{2}\right\}^2} \text{(since } \sqrt{10} = \pi\text{)}$$

Mahāvīra gives the formulae: area of a segment $= \frac{\pi c.\,h}{4}$

(*G. S. S.* VII $70\frac{1}{2}$)

and arc$=\sqrt{6\,h^2+c^2}$ (ibid. VII. $73\frac{1}{2}$)

As the *vyāvahārika* values of these (i.e. approximate values useful for everyday calculations) are given

$$\text{area} = (c+h)\,\frac{h}{2}$$

$$\text{arc} = \sqrt{5\,h^2+c^2}$$

(*G. S. S.* VII. 43)

The only difference between the two sets is that the one is derived with $\pi=\sqrt{10}$, while the other is with $\pi = 3$.

[1]B.B. Datta (*Jaina Antiquary, Bhāga* 2. *Kiraṇa* 2 pp. 34-38) says that V. 47 of the *Triśatikā* embodies the formula, area of segment$=\frac{1}{2}$(c+h) h.

Āryabhaṭa II reproduces the ancient formula for the arc ($\sqrt{6h^2+c^2}$) and the formulae for the height and chord derived from it. His formula for the approximate area of a segment is

ज्याबाणैक्यदलेष्वोघाताात्[1] स्वघ्नात् स्वनवमभागयुतात् ।
यन्मूलं तत् स्थूलं क्षेत्रफलं कार्मुके भवति ॥

(*Ma. Si.* XIV. 89)

(The product of half the sum of the chord and the arrow and the arrow multiplied by itself is combined with one-ninth of itself. The square root of the resulting quantity is the approximate area of the segment)

$$\text{i.e. area of segment} = \sqrt{\left\{\frac{h\,(c+h)}{2}\right\}^2 + \frac{1}{9}\,\frac{h\,(c+h)}{2}}$$

If, instead, *svanavamabhāga* is taken as $\frac{1}{9}\left\{\frac{h\,(c+h)}{2}\right\}^2$, the

$$\text{expression reduces to } \sqrt{\frac{10}{9}\left\{\frac{h\,(c+h)}{2}\right\}^2}$$

$$= \frac{\pi}{3} \cdot \frac{h\,(c+h)}{2}$$

This is Śrīdhara's formula.

Śrīpati omits the expression for the area of the segment. But the whole series for the height, chord and diameter, each in terms of the others, and for the arc, chord and height, each in terms of the rest is given.

Bhāskara II's commentator Sūryadāsa (early 16th century) derives the familiar formulae for the chord, height and diameter in a new way.

[1]Sudhakara Dvivedi accepts the reading ज्याबाणैक्यदलज्याघातात् and the reading accepted here is given in a footnote. With Dvivedi's reading the formula contained in the verse will be area of segment $= \frac{\pi}{3} \cdot c\,\frac{(c+h)}{2}$ which gives a value grossly in excess of the actual area of a segment. Also when giving the exact area (i.e. with $\pi = \frac{22}{7}$) the formula used by Āryabhaṭa is area $= \frac{\pi}{3}\,h.\,\frac{(c+h)}{2}$

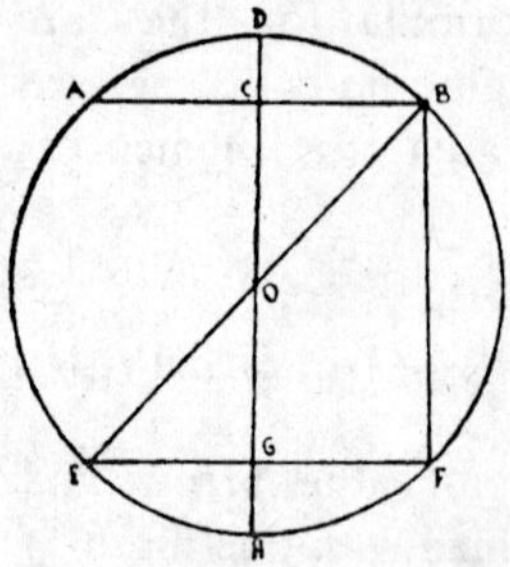

Fig. 10

Let AB be any chord and EF an equal chord vertically opposite to it (i.e. parallel to it). Join EB and BF. Let DH be the diameter bisecting A B at C and E F at G.

$$\text{Then } CD = \frac{DH-CG}{2} = \frac{DH-BF}{2}$$

$$= \frac{DH-\sqrt{BE^2-EF^2}}{2}$$

$$= \frac{d-\sqrt{d^2-c^2}}{2}$$

Bhāskara discards the old crude methods of finding the chord in terms of the arc and vice versa and gives much better approximations, though he recognises these too are approximations only.

चापोननिघ्नपरिधिः प्रथमाह्वयः स्यात्
पंचाहतः परिधिवर्गचतुर्थभागः ।
आद्योनितेन खलु तेन भजेच्चतुर्घ्न—
व्यासाहतं प्रथममाप्तमिह ज्यका स्यात् ॥

(*Līl.* 210)

(The circumference diminished and multiplied by the arc shall be called the first (*prathama*). One quarter of the circumference multiplied by 5 is to be diminished by the *prathama*. The *prathama* multiplied by 4 and the diameter should be divided by the above difference. The quotient will be the chord.)

That is, if c is the chord of the arc a, and if d and C are the diameter and circumference of the circle whose part the arc is

$$c = \frac{4d\,(C-a)a}{5C^2/4-(C-a)\,a}$$

This is, as already pointed out in Ch. I, a modification of Bhāskara I's formula $r.\sin\theta = \dfrac{r.\,\theta\,(180-\theta)}{\frac{1}{4}[40500-\theta(180-\theta)]}$ for computing sines directly from the arcs, and it gives very close approximations.

The converse is

व्यासाब्धिघातयुतमौर्विकया विभक्तो
जीवांघ्रिपंचगुणितः परिधेस्तु वर्गः ।
लब्धोनितात् परिधिवर्गचतुर्थपादात्
आप्ते पदे वृतिदलात् पतिते धनुः स्यात् ॥ (*Lil.* 212)

(The square of the circumference multiplied by 5 and one-fourth of the chord is divided by the chord combined with 4 times the diameter. When the square root of one fourth the square of the circumference diminished by the above quotient is subtracted from the semicircumference, we get the arc.)

With the same symbolism as before this means

$$a = \frac{C}{2} - \sqrt{\frac{C^2}{4} - \frac{5C^2\ c}{4(c+4d)}}$$

Bhāskara's rule for the chord from the arc is reproduced by Nārāyaṇa with a change in form[1]

$$c = d\ \frac{\left\{\left(\frac{C}{2}\right)^2 - \left(\frac{C}{2} - a\right)^2\right\}}{\left\{C^2 + \left(\frac{C}{2} - a\right)^2\right\} \Big/ 4}$$

As the editor of the *Gaṇitakaumudī* shows, this, when simplified, reduces to Bhāskara's expression.

The most significant achievements in approximating to the value of the arc in terms of the chord belong to the Āryabhaṭa school. The simplest of these expressions for the arc is to be credited to Nīlakaṇṭha Somayājin.

सत्त्र्यंशादिषुवर्गाज्ज्यावर्गाढ्यात् पदं धनुः प्रायः ।

(*A. B. Gaṇitapāda, p.* 63)

(The square root of the sum of one and one-third the square of the arrow and the square of the (sine) chord is the arc nearly)

$$\text{i.e. } a = \sqrt{(1 + \tfrac{1}{3})\ h^2 + s^2}$$

Here the arc is calculated in terms of its sine chord (s) and its versed sine chord which is the same as the height of twice the arc.[2] Let h be this height and c_1, c_2, c_3........the *whole* chords of arc AB and of the arcs got by successively halving it. Let h_1,

[1]*G.K. Ks. Vya.*, 69.

[2]*A.B. Gaṇitapāda*, p. 101ff.

h_2, h_3.........be the heights of these arcs. In the figure, AB is the arc required to be found and arc AC is double arc AB. If through B the diameter BOK is drawn, it will bisect the chcrd AC at right angles at D. Then AD is the given sine-chord (*ardhajyā*) of the arc AB

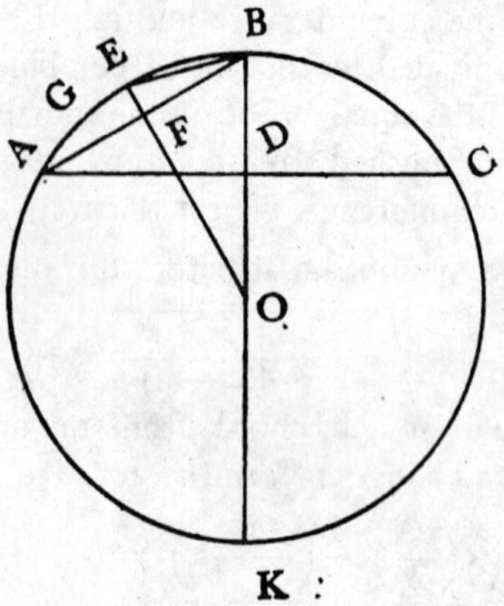

Fig. 11

$$\text{and } c_1^2 = AB^2 = AD^2 + BD^2$$
$$= s^2 + h^2$$

If E, the middle point of arc AB, is joined to O, the centre of the circle, cutting chord A B at F,

$$c_2^2 = AE^2 = AF^2 + EF^2$$
$$= \frac{c_1^2}{4} + h_1^2$$

$$\text{Similarly} \quad c_3^2 = \frac{c_2^2}{4} + h_2^2$$
$$= \frac{c_1^2/4 + h_1^2}{4} + h_2^2$$
$$= \frac{c_1^2}{4^2} + \frac{h_1^2}{4} + h_2^2$$
$$c_4^2 = \frac{c_3^2}{4} + h_3^2$$

$$= \frac{c_1^2}{4^3} + \frac{h_1^2}{4^2} + \frac{h_2^2}{4} + h_3^2$$

. . . .

. . . .

$$c_n^2 = \frac{c_1^2}{4^{n-1}} + \frac{h_1^2}{4^{n-2}} + \frac{h_2^2}{4^{n-3}} + \ldots + h^2_{n-1}$$

If n is sufficiently large the arc will be very small and the chord can be equated to the arc, when we can write

$$a_n^2 = \frac{c_1^2}{4^{n-1}} + \frac{h_1^2}{4^{n-2}} + \frac{h_2^2}{4^{n-3}} + \ldots + h^2_{n-1}$$

Proceeding inversely the square of twice this arc

$$\text{i.e. } a^2_{n-1} = \frac{c_1^2}{4^{n-2}} + \frac{h_1^2}{4^{n-3}} + \frac{h_2^2}{4^{n-4}} + \ldots + 4\, h^2_{n-1}$$

$$a^2_{n-2} = \frac{c_1^2}{4^{n-3}} + \frac{h_1^2}{4^{n-4}} + \frac{h_2^2}{4^{n-5}} + \ldots + 4^2\, h^2_{n-1}$$

. . . .

. . . .

$$a_2^2 = \frac{c_1^2}{4} + h_1^2 + 4\, h_2^2 + \ldots + 4^{n-2} \cdot h^2_{n-1}$$

$$a_1^2 = c_1^2 + 4\, h_1^2 + 4^2\, h_2^2 + \ldots + 4^{n-1}\, h^2_{n-1}$$

$$= s^2 + 4\, h_1^2 + 4^2\, h_2^2 + \ldots + 4^{n-1}\, h^2_{n-1}$$

Now $BD = h = \frac{AB^2}{BK} = \frac{c_1^2}{d}$ (from the similar triangles BAD and BAK)

$$EF = h_1 = \frac{AE_2^2}{d}$$

$$= \frac{\frac{AB^2}{4} + EF^2}{d}$$

$$= \frac{c_1^2}{4d} + \frac{h_1^2}{d}$$

$$= \frac{c_1^2}{4d} \text{ (neglecting higher powers of } h_1 \text{ etc.)}$$

$$= \frac{h}{4}$$

Similary, $h_2 = \frac{h_1}{4} = \frac{h}{16} = \frac{h}{4^2}$

$$h_3 = \frac{h}{4^3}$$

Therefore $h^2, h_1^2, h_2^2 \ldots\ldots \quad \ldots\ldots$ will be a geometrical progression with $\frac{1}{16}$ as common ratio.

Hence $h^2 + 4\,h_1^2 + 4^2\,h_2^2 + \ldots\ldots \quad \ldots\ldots$

$$= h^2 + \frac{h^2}{4} + \frac{h^2}{4^2} + \frac{h^2}{4^3} + \ldots\ldots \quad \ldots\ldots$$

$$= h^2\left(1 + \frac{1}{4} + \frac{1}{4^2}\right) + \ldots\ldots \quad \ldots\ldots$$

$$= h^2\left(\frac{1}{1-\frac{1}{4}}\right) = h^2 \, . \, 1\frac{3}{4} = \frac{4}{3}h^2$$

$$\therefore\ a_1^2 = s^2 + h^2\left(1 + \tfrac{1}{3}\right)$$

Nīlakaṇṭha notes that this is only an approximation, since in the calculation of h_1, h_2 etc. their powers are neglected. But since the neglected powers are divided by the diameter also, these will be negligibly small for small arcs, and Nīlakaṇṭha recommends the use of the formula for small arcs only. Again he insists that all this is implied in Āryabhaṭa's *sūtra* वृत्ते शरसंवर्गः अर्धज्यावर्गः स खलु धनुषोः ।

7.11. *Mādhava s discovery in the* 14*th century of Gregory's series*

As for the circumference in terms of the diameter, this school has to its credit an expression for the arc in the form of an infinite series in powers of the sine-chord and cosine-chord, which is capable of being wielded to yield any desired degree of closeness. The series is embodied in the following verses quoted in the *Kriyākramakarī* and assigned to Mādhava[1] (14th century).

इष्टज्यात्रिज्ययोर्घातात् कोट्याप्तं प्रथमं फलं ।
ज्यावर्गं गुणकं कृत्वा कोटिवर्गं च हारकम् ॥
प्रथमादिफलेभ्योऽथ नेया फलततिर्मुहुः ।
एकत्र्याद्योजसंख्याभिर्भक्तेष्वेतेष्वनुक्रमात् ।
ओजानां संयुतेस्त्यक्त्वा युग्मयोगं धनुर्भवेत् ।
दोःकोट्योरल्पमेवेष्टं कल्पनीयमिह स्मृतम् ॥ *K.K.* 692-693

(The product of the given sine-chord and the radius, divided by the cosine chord, is the first result. Then a series of results

[1]Also quoted in the *Y.B.* (p. 113) and assigned to the *Tantrasaṁgraha*, which, it is to be remembered, takes much from Mādhava.

are to be obtained from this first result and the succeeding ones by making the square of the sine-chord the multiplier and the square of the cosine-chord the divisor. When these are divided in order by the odd numbers 1, 3, etc., the sum of the terms in the even places is to subtracted from the sum of the terms in the odd places to get the arc. The smaller of the sine and cosine-chords is to be used for this calculation.)

That is if s and c are the sine and cosine chords.

$$\text{the arc} = \frac{s\,r}{c} - \frac{s\,r}{3\,c}\cdot\frac{s^2}{c^2} + \frac{s\,r}{5\,c}\cdot\frac{s^4}{c^4} - \frac{s\,r}{7\,c}\cdot\frac{s^6}{c^6} + \ldots\ldots$$

$$\text{i.e. arc} = \frac{s}{c}\,r - \frac{1}{3}\cdot\frac{s^3}{c^3}\cdot r + \frac{1}{5}\cdot\frac{s^5}{c^5}\cdot r - \frac{1}{7}\cdot\frac{s^7}{c^7}\cdot r + \ldots$$

Since $\frac{s}{c}$ = tangent of the same arc, this equality, in a circle of unit radius reduces to arc $= t - \frac{1}{3}t^3 + \frac{1}{5}t^5 - \frac{1}{7}t^7 + \ldots\ldots\ldots$ where $t = \tan\theta$. And this is the series rediscovered in Europe by James Gregory in 1671.

The method and principle of derivation are the same as those for finding the circumference in terms of the radius.

In the course of deriving the series for π, it was shown that

$$\frac{\text{circumference}}{8} = R - \frac{R^3}{3R^2} + \frac{R^5}{5R^4} - \frac{R^7}{7R^6} + \ldots\ldots$$

Here instead of evaluating $\frac{1}{8}$ the circumference, we have to evaluate the arc whose sine-chord is s.

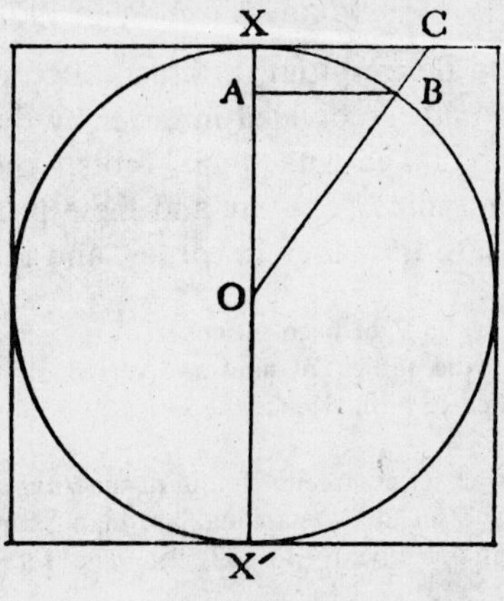

Fig. 12.

Let AB be the half-chord = s in the circle with centre O inscribed in a square of side=2r. As for finding $\frac{1}{8}$ the circumference, join O B and produce it to cut the side of the square in C. Then the number of terms in the infinite series summed up, will be the number of unit divisions in XC. Hence the expression for the arc is arc

$$= XC - \frac{XC^3}{3\,r^2} + \frac{XC^5}{5r^4} - \ldots$$

Now from the similar triangles XOC and AOB

$$\frac{XC}{AB}=\frac{OX}{OA} \text{ or } XC=\frac{AB \cdot OX}{OA}=\frac{s \cdot r}{c.}$$

Hence, substituting in the series,

$$\text{arc} = \frac{s \cdot r}{c}-\left(\frac{s \cdot r}{c}\right)^3 \cdot \frac{1}{3\,r^2}+\left(\frac{s \cdot r}{c}\right)^5 \cdot \frac{1}{5\,r^4}- \ldots\ldots\ldots$$

$$= \frac{s \cdot r}{c}-\frac{s \cdot r}{3c} \cdot \frac{s^2}{c^2}+\frac{s \cdot r}{5c} \cdot \frac{s^4}{c^4}- \ldots\ldots\ldots$$

By taking chords of known arcs (as for example the chord of $\frac{1}{12}$ circumference (30°) $=\frac{r}{2}$) we can get other series for π from this. It is obvious that arcs above $\frac{1}{8}$ the circumference cannot be evaluated by this method.

7.12. *The Sine and Cosine series*

The sine and cosine series well-known in the Āryabhaṭa School and derived in the *Yukti Bhāṣa*, are two more of the achievements of the great Mādhava.[1] For the sine series the *Tantrasaṃgraha* has

निहत्य चापवर्गेण चापं तत्तत् फलानि च।
हरेत् समूलयुग्वर्गैस्त्रिज्यावर्गाहतैः क्रमात्।
चाँप फलानिं चाधोऽधौ न्यस्योपर्युपरि त्यजेत्।
जीवाप्त्यै ....

(quoted in *Y.B.* p. 190)[2]

(One should multiply the arc and the resulting products by the square of the arc, and these should be divided in order by the squares of the even numbers combined with their roots, and multiplied by the square of the radius. The arc and these products should be placed one beneath the other in order and the

[1]Verses embodying the two series have not yet been discovered to occur as quotations from Mādhava. But the tables of sines and versed sines calculated with their help are attributed to him. Hence the series also must have been discovered by him.

[2]The verses for the sine and cosine series are not found in the *Tantrasaṃgraha* published from Trivandrum. But the Desamangalam manuscript with a Malayalam commentary contains them (pp. 58-59). The series occur in the *Karaṇapaddhati* too (VI 12, 13).

lower ones should be subtracted from the upper ones for getting the sine-chord.)

The various terms are, if a is the arc, and r the radius,

$$\frac{a \,.\, a^2}{(2^2+2)r^2} = \frac{a^3}{\lfloor 3 r^2}$$

$$\frac{a^3 \,.\, a^2}{\angle 3(4^2+4)r^4} = \frac{a^5}{\angle 5 \,.\, r^4}$$

$$\frac{a^5 \,.\, a^2}{\angle 5(6^2+6)r^6} = \frac{a^7}{\angle 7 \,.\, r^6}$$

$$\text{Hence sine-chord} = a - \frac{a^3}{\angle 3 \,.\, r^2} + \frac{a^5}{\angle 5 \,.\, r^4} - \frac{a^7}{\angle 7 \,.\, r^6} + \ldots$$

The series for the versed sine is made similarly

निहत्य चापवर्गेण रूपं तत्तत् फलानि च ।
हरेत् विमूलयुग्वर्गैस्त्रिज्यावर्गाहतैः क्रमात् ॥
किन्तु व्यासदलेनैव द्विघ्नेनाद्यं विभज्यताम् ।
फलान्यधोऽधःक्रमशो न्यस्योपर्युपरि त्यजेत् ॥
शराप्त्यै.......

(One should multiply one and the resultant products by the square of the arc and divide in order by the squares of the even numbers diminished by their roots, and multiplied by the square of the radius. But let the first be divided by the radius multiplied by two only. These terms should be placed one beneath the other in order and the lower ones should be subtracted in order from the one above them in order to get the arrow....)

i.e. the versed sine or the *śara* of the sine chord, as the Indians put it

$$= \frac{a^2}{2.r} - \frac{a^4}{2 \,.\, r^3 \,.\, 12} + \frac{a^6}{2 \,.\, 12 \,.\, r^5 \,.\, 30} - \ldots$$

$$= \frac{a^2}{\angle 2 \,.\, r} - \frac{a^4}{\angle 4 \,.\, r^3} + \frac{a^6}{\angle 6 \,.\, r^5} - \ldots$$

The derivation of these series making use of the *saṃkalita* of the parts of a whole or integration employed in getting the series for π, is based on still subtler analysis and proceeds stage by stage. The first step is to find expressions for the change in the sine-chord (*bhujajyā*) and the cosine-chord (*koṭijyā*) produced by a small difference in the arc, that is to differentiate the sine and the cosine.

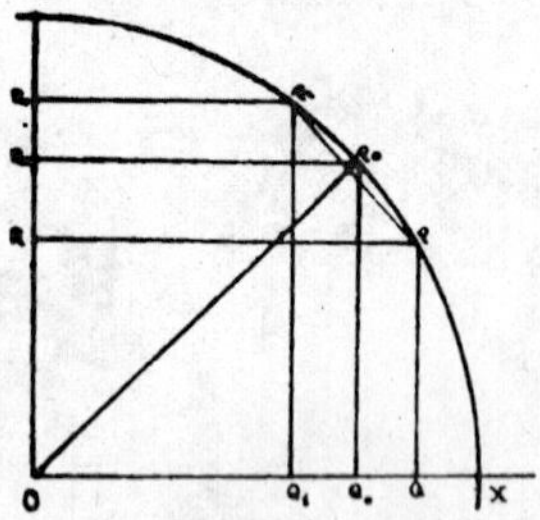

Fig. 13

Let P P_1 represent a small increase in the arc PX. Draw the *bhujajyā* and *koṭijyā* (P_1Q_1 and P_1R_1) at P_1 and also those at the middle point of the small arc (P_0Q_0 and P_0R_0). Then, if the point of intersection of P_1Q_1 and PR is S, triangles P_1SP and OQ_0P_0 are similar —

$$\therefore \frac{P_1S}{P_1P} = \frac{OQ_0}{OP_0}$$

$$\text{or } P_1S = \frac{OQ_0 \cdot P_1P}{OP_0}$$

$$\text{and } P\,S = \frac{P_0Q_0 \cdot P_1P}{OP_0}$$

But if the arc PP_1 is very small, the sine-chord and cosine-chord at P_0 may be taken to be equal to those at P and the arc PP_1 to coincide with the chord P P_1.

$\therefore$ P_1 S (sine difference or *bhujākhaṇḍa*)

$$= \frac{\text{the cosine chord at P . the increase in the arc}}{\text{radius}}$$

i.e. in modern notation, $d\,(r \sin a) = \dfrac{r \cos a \,.\, d\,a}{r}$

Similarly the *koṭikhaṇḍa* $d\,(r. \cos a) = \dfrac{r \sin a \,.\, d\,a}{r}$

This theorem is stated and proved both in the *Yukti-bhāṣā* (pp. 165 & 173) and in Nīlakaṇṭha's *Āryabhaṭīya-bhāṣya* (*Gaṇita-pāda* pp. 48-50) as an elaboration of one of Mādhava's rules. The result was known to Bhāskara II also.

By applying the principle of proportionality once more, the *khaṇḍāntaras* or the second differences of the sine and cosine chords, can be calculated. For the difference between any two *bhujākhaṇḍas*, i.e. the second sine difference

$$= \text{2nd } kotijy\bar{a} \cdot \frac{d\,a}{r} - \text{1st } kotijy\bar{a} \ \frac{d\,a}{r}$$

$$= \frac{d\,a}{r} . kotikhanda$$

$$= \frac{d\,a}{r}\left(bhujajy\bar{a} \cdot \frac{d\,a}{r}\right)$$

$$= bhujajy\bar{a} \cdot \frac{(d\,a)^2}{r^2}$$

$$= r \cdot \sin a \cdot \frac{(d\,a)^2}{r^2}$$

This result also occurs both in the *Āryabhaṭīya-bhāṣya* (pp. 51-53) and in the *Yuktibhāṣā* (p. 173).

The above result is used in the *Yuktibhāṣā* to find an expression for the difference between any arc and its sine chord. Let the arc be marked off in the first quadrant and divided into n equal parts ($\triangle a$). Let s_1, s_2, s_3......be the sine chords at these points and $\triangle s_1$, $\triangle s_2$......the sine differences at these points.

Then clearly $\triangle s_1 = s_1$

$$\triangle s_1 - \triangle s_2 = s_1 \cdot \frac{(d\,a)^2}{r^2}$$

$$\text{or} \quad \triangle s_2 = \triangle s_1 - s_1 \cdot \frac{(d\,a)^2}{r^2}$$

$$= s_1 - s_1 \cdot \frac{(d\,a)^2}{r^2}$$

Similarly $\triangle s_3 = \triangle s_2 = s_2 \frac{(d\,a)^2}{r^2} = s_1 - s_1 \cdot \frac{(d\,a)^2}{r^2} - s_2 \frac{(d\,a)^2}{r^2}$

... ...

... ...

$$\therefore s_n = s_1 - s_1 \cdot \frac{(d\,a)^2}{r^2} - s_2 \cdot \frac{(d\,a)^2}{r^2} \ldots \ldots s_{n-1} \frac{(d\,a)^2}{r^2}$$

$\therefore s_n$, the sum of all these sine differences =

$$n.s_1 - \left\{(n-1)s_1 + (n-2)s_2 + \ldots\ldots + 2\,s_{n-2} + s_{n-1}\right\}\frac{(d\,a)^2}{r^2}$$

If n is sufficiently large, the first chord will nearly coincide with the arc. Hence $n.s_1$ = the whole arc.

∴ The whole arc $-s_n$ i.e. the difference between the arc and the sine chord $= \left\{(n-1)\, s_1 + (n-2)s_2 + \ldots \ldots \ldots s_{n-1}\right\} \frac{(d\,a)^2}{r^2}$

This is also equal to the *saṃkalita* of the second sine differences, Now the height ((*śara*) of the given arc = the sum of the cosine differences $= s_1 \cdot \frac{d\,a}{r} + s_2 \cdot \frac{d\,a}{r} + \ldots \ldots$

$$= \frac{d\,a}{r}(s_1 + s_2 + \ldots \ldots)$$

These unknown sine-chords are again assumed to be equal to the arcs themselves and d a to be equal to one unit.

Then the height of the arc $= \frac{1}{r}(1+2+3+ \ldots \ldots + a)$

$$= \frac{1}{r} \cdot \frac{a^2}{2}$$

The sine differences $\triangle s_1, \triangle s_2 \ldots,$ and hence $(\triangle s_1 - \triangle s_{n-1}) \ldots$ can be expressed in terms of the corresponding arcs.

For $\triangle s_1 = c_{\frac{1}{2}} \cdot \frac{da}{r}$ where $c_{\frac{1}{2}}$ is the cosine chord at the middle point of the first arc bit,)

$= \frac{da}{r}(r - h_{\frac{1}{2}})$ ($h_{\frac{1}{2}}$ stands for the height of the half-arc)

Similarly $\triangle s_2 = c_{1\frac{1}{2}} \cdot \frac{da}{r} = (r - h_{1\frac{1}{2}})\frac{da}{r}$

... ...

... ...

$$\triangle s_n = c_{n-\frac{1}{2}} \cdot \frac{da}{r} = (r - h_{n-\frac{1}{2}})\frac{da}{r}$$

$$\therefore \triangle s_1 - \triangle s_n = \frac{d\,a}{r}(r - h_{\frac{1}{2}} - r + h_{n-\frac{1}{2}})$$

Since da is very small $h_{\frac{1}{2}}$ can be deemed to be equal to $h_0 = 0$, and $h_{n-\frac{1}{2}} = h_n$

$$\therefore \triangle s_1 - \triangle s_n = \frac{d\,a}{r} \cdot h_n$$

$$= \frac{d\,a}{r} \cdot \frac{a^2}{2\,r}$$

$$= \frac{a^2}{2\,r^2} \text{ if } da = 1 \text{ unit}$$

Similarly $\triangle s_1 - \triangle s_{n-1} = \dfrac{(a-1)^2}{2\,r^2}$

$$\triangle s_1 - \triangle s_{n-2} = \frac{(a-2)^2}{2\,r^2}$$

$$\ldots \qquad \ldots \qquad \ldots$$

$$\ldots \qquad \ldots \qquad \ldots$$

The sum of all these = the *saṃkalita* of the second sine differences

$$\frac{a^2}{2\,r^2} + \frac{(a-1)^2}{2\,r^2} + \ldots$$

$$= \frac{1}{2r^2} \cdot \frac{a^3}{3} = \frac{a^3}{r^2 . \angle 3}$$

But the *saṃkalita* of the 2nd sine differences has already been shown to be equal to the difference between the arc and its sine-chord.

$$\therefore \text{Arc—sine chord} = \frac{a^2}{2\,r^2} + \frac{(a-1)^2}{2\,r^2} + \ldots \qquad \ldots$$

$$= \frac{1}{2\,r^2} \cdot \frac{a^3}{3} = \frac{a^3}{r^2 . \angle 3}$$

Here, since, in the derivation of the expression for the height of the arc, the sine chords have been equated to the arcs to which they are not really equal, this same difference between the arc and its sine-chord has to be applied as correction to the arcs appearing in the expression, i.e. in

$$\frac{1}{r}\left\{a + (a-1) + (a-2) + \ldots \ldots\right\}$$

When the correction is applied this becomes

$$\frac{1}{r}\left\{(a - \frac{a^3}{\angle 3 . r^2}) + (a - 1 - \frac{(a-1)^3}{\angle 3 . r^2} + \ldots\ldots\right\}$$

$$= \frac{1}{r}\left\{a + (a-1) + \ldots \right\} - \frac{1}{r}\left\{(\frac{a^3}{\angle 3 . r^2} + \frac{(a-1)^3}{\angle 3 . r^2} + \ldots \right\}$$

Hence the correction for the height

$$= \frac{1}{r}\left\{\frac{a^3}{\angle 3 \,.\, r^2} + \frac{(a-1)^3}{\angle 3 \,.\, r^2} + \dots \dots \dots\right\}$$

$$= \frac{1}{\angle 3 \,.\, r^3} \,.\, \frac{a^4}{4} = \frac{a^4}{r^3 \,.\, \angle 4}$$

When this corrected value is used in the expression for the arc-sine difference, it expands to

$$\frac{1}{r}\left\{\frac{a^2}{2r} + \frac{(a-1)^2}{2\,r} + \dots \dots\right\} - \frac{1}{r}\left\{\frac{a^4}{r^3 \,.\, \angle 4} + \frac{(a-1)^4}{r^3 \,.\, \angle 4} + \dots\right\}$$

$$= \frac{a^3}{2\,r^2 \,.\, 3} - \frac{a^5}{r^4 \,.\, \angle 4 \,.\, 5}$$

$$= \frac{a^3}{\angle 3 \,.\, r^2} - \frac{a^5}{r^4 \,.\, \angle 5}\,.$$

Thus, these two corrections can be applied *ad infinitum* to the interdependent expressions for the arc-sine difference and the height. Hence finally the *jyā* or sine-chord

$$= a - \frac{a^3}{r^2 \angle 3} + \frac{a^5}{r^4 \angle 5} - \frac{a^7}{r^6 \,.\, \angle 7} + \dots \dots$$

And the cosine-chord = r — the height of the arc

$$= r - \frac{a^2}{r \,.\, \angle 2} + \frac{a^4}{r^3 \,.\, \angle 4} - \frac{a^6}{r^5 \,.\, \angle 6} + \dots$$

Putting $a = r\,\theta$ and remembering that the sine and cosine chords are $r \sin \theta$ and $r, \cos \theta$, we get the series

$$\text{Sin } \theta = \theta - \frac{\theta^3}{\angle 3} + \frac{\theta^5}{\angle 5} - \frac{\theta^7}{\angle 7} + \dots$$

$$\text{Cos } \theta = 1 - \frac{\theta^2}{\angle 2} + \frac{\theta^4}{\angle 4} - \frac{\theta^6}{\angle 6} + \dots\dots$$

Series for trigonometrical functions were known in Europe by the 17th century, whereas in India they were known in the 14th century.

7.12. *The common chord and its height in intersecting circles*

Problems connected with intersecting circles were important for the calculation of eclipses and hence they find a place in most Indian mathematical and astronomical works. Āryabhaṭa's rule for calculating the heights of the arcs enclosing the common portion of the circles is

ग्रासोने द्वे वृत्ते ग्रासगुणे भाजयेत् पृथक्त्वेन ।
ग्रासोनयोगलब्धी सम्पातशरौ परस्परतः ॥

(*A. B. Gaṇitapāda*, 18)

(The two diameters diminished by the *grāsa* (the largest breadth of the common portion) and multiplied by the *grāsa* should be divided by the sum of the diameters less the *grāsa*. These will be the height of the arcs of the circles in the common portion, each of the other.)

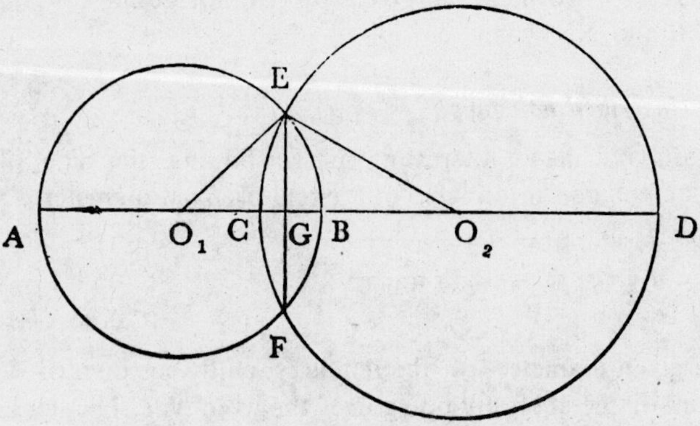

Fig. 14

The circles with centres O_1 and O_2 interesect at E and F. The line joining their centres cut the circumferences at A, B, and C, D. Then C B is the *grāsa*. If EF, the common chord, cuts the line of centres in G, GB and GC are the arrows and the rule tells us how to calculate these. If d_1 and d_2 are the diameters and h_1 and h_2 the heights of the arcs

$$h_1 = \frac{\{d_2-(h_1+h_2)\}\ (h_1+h_2)}{d_1+d_2-2\ (h_1+h_2)}$$

$$\text{and } h_2 = \frac{\{d_1-(h_1+h_2)\}\ (h_1+h_2)}{d_1+d_2-2\ (h_1+h_2)}$$

This is derived by Nilakaṇṭha using Āryabhaṭa's expression for the chord of a segment in terms of the diameter and height. Here since E F is a chord common to the two circles

$$\frac{EF^2}{4} = h_1 (d_1 - h_1) = h_2 (d_2 - h_2).$$

Hence $\dfrac{h_1}{h_2} = \dfrac{d_2 - h_2}{d_1 - h_1}$

$$= \frac{d_2 - h_2 - h_1}{d_1 - h_1 - h_2}$$

$$\therefore \frac{h_1 + h_2}{h_2} = \frac{d_2 + d_1 - 2(h_2 + h_1)}{d_1 - (h_1 + h_2)}$$

$$\therefore h_2 = \frac{\{d_1 - (h_1 + h_2)\}(h_1 + h_2)}{d_1 + d_2 - 2(h_2 + h_1)}$$

The same formula is given by Brahmagupta, Mahāvīra, Śrīpati and Nārāyaṇa.

7.13. *Inscribed polygons*

Mahāvīra has an interesting rule for finding the side of any regular polygon inscribable in a circle of given diameter.

लब्धव्यासेनेष्टव्यासो वृत्तस्य तस्य भक्तश्च ।
लब्धेन भुजा गुणयेत् भवेच्च जातस्य भुजसंख्या ॥

(*G.S.S.* VII. 221½)

(The given diameter of the circle should be divided by the circum-diameter of any polygon of the given type. The sides of the polygon multiplied by this ratio will be the sides of the required polygon (i.e. the similar polygon inscribable in the given circle.) The method can be used to inscribe rational polygons in a given circle, provided the method for forming a rational polygon of the given type is known. Thus from the formula for the construction of rational triangles and cyclic quadrilaterals, similar rational figures can be inscribed in the given circle.[1]

Mahāvīra calculates the diagonals and altitudes of a regular hexagon, but gives a wrong formula for its area (VII. 86½).

Bhāskara gives the sides of regular polygons of sides 3, 4, 5, 6, 7, 8, 9, inscribed in a circle of diameter 120000. We do not

[4]Dr. Datta (*On Mahavira's solution of Rational Triangles and Quadrilaterals Bull. Cal. Math. Soc.* XX pp. 291-92) points out that the problem of inscribing rational polygons in a given circle was attempted by Euler (c. 1781) and H. Schubert (1905).

know what his method for getting these was. Gaṇeśa divides the circumference into as many equal parts and finds the chord corresponding to one division using the sine table. But the results do not tally exactly with Bhāskara's results. Alternatively he calculates the sides of the triangle, the square, the hexagon and the octagon geometrically.

Nārāyaṇa's method is the first one

ज्यापरिधिरश्मिभागाद् धनुरथवा रश्मिसम्मितः परिधिः ।
रूपं चापं तज्ज्या तुल्यत्र्यस्रादिभुजमानम् ॥

(*G. K. Ks. Vya.* 72)

(The arc is to be obtained from the circumference divided by the number of sides. The chord of this arc will be the side of the (inscribed) regular triangle and other polygons. Or let the circumference be equal to the number of sides. Then the arc is one unit. Its chord is the measure of the side of the regular triangle and other figures.)

In the latter method the actual sides are to be obtained by multiplying the chords by the ratio of the diameter of the given circle to the diameter of the circle used for the construction.

An interesting calculation that Mahāvīra does, is finding the number of small cylinders in a cylindrical container, when the diameter of the small cylinders and their number in the outermost layer are known. The problem is connected with the arrangement of arrows in a quiver and the solution is : "The number of arrows, forming the circumferential layer, combined with 3, squared and again combined with 3 is to be divided by 12. The quotient will be the number of arrows in the quiver" (*G. S. S.* VI. 288) i.e., if n is the number in the outermost layer, the

total number $N = \frac{(n+3)^2+3}{12}$

The formula is based on the knowledge that only six equal circles can be drawn around a circle of equal radius touching it and each other, which, in its turn, is based on the fact that the lines joining the centres of 3 equal circles touching one another is an equilateral triangle.[1] Then the number of circles in the succeeding layers will be 2×6, 3×6 etc.

[1]The rationale is given by Prof. Rangacharya.

Fig. 15

$\therefore$ The number of circles in p layers (along with the central circle $=1+6+2.6+\ldots\ldots\times p.6=1+6\,(1+2+\ldots\ldots+p)$

$$=1+\frac{6\,p\,(p+1)}{2}=1+3\,p\,(p+1).$$ If the pth layer contains

n circles, $p=\frac{n}{6}$, so, $N=1+\frac{3n}{6}\left(\frac{n}{6}+1\right)=1+\frac{3n^2}{36}+\frac{3n}{6}$

$$=\frac{12+n^2+6n}{12}=\frac{(n+3)^2+3}{12}$$

Chapter VIII

VOLUMES AND SURFACES OF SOLIDS

8.1. The *Śulbasūtras* do not speak of volumes directly. But they must have been familiar with the concept, since they used to fix the height as well as the number of layers and total number of bricks in the fire altars of regular geometrical shape. They might also have realised that the volumes of regular solids with opposite faces parallel is the product of the area of the base and the height.

8.2. We have indisputable evidence of their knowledge that a parallelopiped is cut into two prisms of equal volume by the plane passing through the parallel diagonals of one pair of opposite faces. Baudhāyana enjoins that the *śmaśānaciti* fireplace should have for base an isosceles trapezium and that its top surface should slope from one edge (eastern) to the other (western) so that its eastern height is up to the neck and the western height is up to the navel.[1] And yet the volume of the fire-place is to be the same as that of the usual fireplace. The method adopted to achieve this is:[2] The usual vertical height of the fire altar is increased by one-fifth of itself. Then this whole height is divided into 3 parts. The fire altar is built up in the usual way and the top one third is then sliced diagonally by a plane passing through the top eastern edge and reaching up to the bottom western edge of the one third part. Then since the third part is $\frac{6}{5} \cdot \frac{1}{3}$ i.e. $\frac{2}{5}$ of the whole fireplace, what remains after slicing is $\frac{2}{3} \cdot \frac{6}{5} + \frac{1}{2} \cdot \frac{2}{5} = 1$ and the added one-fifth is removed.[3] The resulting volume will be exactly equal to that

[1] *B.Sr.* xvii. 30 quoted by Dr. Datta on p. 102 of his *Science of the Śulba.*

[2] *B.Sl.* III. 266-269.

[3]. Dr. Datta says (*Science of the Śulba* p. 103) that this construction is based on the approximate formula for the frustum of a pyramid namely,

$$V = \frac{a+a')}{2} \frac{(b+b')}{2} h.$$ But this seems far-fetched. The solids produced by the slicing are prisms or wedges and the whole block also can hardly be called the frustum of a pyramid.

of the usual fire-altar तस्य नित्यो विभाग: (The division of it (the third part) is exact).

It is hard to decide how old the Jaina formulae for volumes are. The *Prajñāpanopāṅga*, (c. 92 B.C.), the *Bhagavatisūtra* and some *karaṇa gāthās* which speak of arranging atoms or shots in the shapes of various plane and solid geometrical figures might have been familiar with the computation of the volumes of such figures. Vīrasena's *Dhavalā* (9th century) and the *Tiloyapaṇṇatti* (c. 8th century), which probably are representatives of a much more ancient tradition, are quite familiar with these formulae. Vīrasena quotes an old *karaṇagāthā* for the volume of a trapezoidal solid.

मुहतलसमास-श्रद्धं वुस्सेधगुणं गुणं च वेधेण ।
घणगणिदं जाणिज्जो वेत्तासणसंठिए खेत्ते ॥

(*Ṣaṭkhaṇḍāgama*, part IV. p. 20)

(Half the sum of the face and the base multiplied by the height and by the depth is to be known as the volume of a figure resembling a rattan seat.)[1] The first part of the rule gives the area of a trapezium and that multiplied by the thickness is the volume of a solid whose section is a trapezium.

i.e. Volume of a trapezoidal solid $= \frac{\text{face}+\text{base}}{2}$.height. depth.

Other Jaina works repeat this formula. The *Tiloyapaṇṇatti* has formulas for computing the volumes of other regular solids also.

1. The volume of the column of air underneath the earth, which is rectangular in section, is given as the product of the length, breadth and thickness (*T.P.* p. 46)

1 मुखतलसमासार्द्धं उत्सेधगुणं च वेधेन ।
घनगणितं जानीयात् वेत्रासनसंस्थिते क्षेत्रे ॥

A separate expression for a figure resembling a *mṛdaṅga* (i.e. two trapezoidal solids joined at their largest face) is given just after this.

मूलं मज्भेण गुणं मुहसहिदद्धमुस्सेधकदिगुणिदम् ।
घणगुणिदं जाणिज्यो मुइगसंठाणखेत्तम्भि ॥

which when literally translated reads "The base multiplied by the middle, combined with the face, halved and multiplied by the square of the height should be known as the volume in a figure resembling a *mṛdaṅga* and does not seem to give any sensible meaning, Maybe the *gāthā* requires emendation.

2. The volume of a cylinder = area of section. height. Though no special rule is given for this calculation, the knowledge is implied in the calculation of the volumes of the imaginary *palyas*[1] (pits) employed by the Jainas to measure very long periods of time.
3. The volume[2] of a prism with right triangular section =

$$\frac{\text{product of perpendicular sides}}{2} \times \text{height}$$

8.3. *The Pyramid*

Āryabhaṭa I seems to give formulae for the volumes of the tetrahedron (a pyramid on a triangular base with all the edges equal) and a sphere, and curiously enough both are wrong. What is more remarkable is that a keen-witted astronomer-mathematician like Nīlakaṇṭha, grown old in the traditions of a school which had discovered the fine tool of differentiation and integration and actually used it to find the volume and surface area of a sphere, should accept these results without a demur and even try to justify them. The one for the tetrahedron is linked with Āryabhaṭa's controversial formula for the area of a triangle.

त्रिभुजस्य फलशरीरं समदलकोटीभुजार्धसंवर्गः।
ऊर्ध्वभुजातत्संवर्गार्धं स घनः षडश्रिरिति॥

(*A.B. Gaṇitapāda* 6)

(The area of a triangle is half the product of the altitude and the base. The product of that and the perpendicular height when halved gives the volume of the 6-edged solid.)

Thus translated, the verse gives a wrong formula for the volume of a pyramid.

$$V = \frac{\text{area of base} \times \text{height}}{2}$$

Nīlakaṇṭha takes *ṣaḍaśri* to mean a pyramid whose faces are equal equilateral triangles. He has also a long note on how to find the length of the *ūrddhvabhuja* (the perpendicular height),

[1] *T.P.* I. 118.

[2] *T.P.* I. 181. The interpretation I give to this verse is different from the one given to it by the editor of the *T.P.*

for which first the circum-centre-cum-ortho-centre of the face and its altitude have to be found. Then, since the line joining the circumcentre to the middle point of the side of the base is $\frac{1}{3}$ the altitude, and since the altitude is $\sqrt{\frac{3a^2}{4}}$ (where a is the edge).

$$\text{the } \bar{u}rdhvabhuja = \sqrt{\frac{3}{4}a^2 - \left(\frac{1}{3}\sqrt{\frac{3}{4}a^2}\right)^2}$$

$$= \sqrt{\frac{3}{4}a^2 - \frac{1}{9}\ \frac{3}{4}a^2}$$

$$= \sqrt{\frac{8}{12}a^2} = \sqrt{\frac{2}{3}a^2}$$

∴ The volume of the tetrahedron

$$= \frac{\frac{1}{2}a.\ \frac{\sqrt{3}\ a.}{2}\ \frac{\sqrt{2}}{\sqrt{3}}.a}{2}$$

$$= \frac{\sqrt{2}\ a^3}{8}$$

The correct value will be $\frac{\sqrt{2}}{12}a^3$

Nīlakaṇṭha here anticipates a question why the 9-edged prism was not dealt with by Āryabhaṭa before the six-edged tetrahedron and answers it by saying

घनन्यायसिद्धत्वात् तस्य । समनवाश्रस्यायतनवाश्रस्य वा तत्संबन्धित्र्यश्रफलस्य तदुच्छ्रितिहननेनैव घनफलं स्यादित्येवं न्यायसिद्धम् । तस्मादूर्ध्वभुजातत्सवर्गं एव नवाश्रफलं इत्येतच्चाप्यनेनैव सूचितम् । ऊर्ध्वभुजातत्संवर्गशब्दोच्चारणात् तत्र बुद्धिः प्रथमं प्रसरेत् । तस्यार्धस्यापनीतत्वात् अवशिष्टं षडश्रक्षेत्रमपि अर्धतुल्यमेवेति भावः ॥

(Because it is implied in the formula for a cube. From this formula, it is clear that the volume of an equilateral prism or of a long prism is obtained by multiplying the area of the triangle connected with it by its height. Hence *ūrddhvabhujātatsamvarga* is the volume of a nine-edged solid (prism). This is indicated by this same *sūtra*. By the utterance of the word *ūrddhvabhujātatsamvarga* the intellect will move towards that first. Since its half is removed, the remaining 6-edged solid also is equal to half of it. This is the idea). And this from a teacher with a fine critical imagination, always prepared to give practical demonstrations with solid figures to support his explanations.

As many have already remarked this mistaken formula for the volume of a pyramid is incompatible with Āryabhaṭa's knowledge of the correct formula for the number of shots in a triangular pile (Gaṇita-pāda 21). Kurt Elfering in his article '*Uber den Flechen - BZW - Rauminalt von Dreieck und Pyramide Sowie Kreis und Kugel bei Aryabhaṭa I*, published in the *Rechenpfennige* (pp 57-68) presented to Kurt Vogel on his 80th Birthday in 1968, re-interprets this verse so as to save Āryabhaṭa from the charge of giving a wrong formula. According to this interpretation the verse should be translated as 'The area of the triangle forming the body or surface (of the pyramid) is half the product of the altitude-bisector and the base, and the solid equal in volume to half the product of that and the height is equivalent to 6 pyramids.'

The pyramid according to Elfering is made by dividing an equilateral triangle into 4 equal equilateral triangles and folding up the peripheral 3 triangles over the central one so that the total surface area of the pyramid = the area of the original triangle. 6 such pyramids together will have volume equal to ½ the product of the area of the original triangle and the height of the pyramid.

For the base area of such a pyramid

$$= \frac{\text{area of original triangle}}{4}$$

and the volume of 6 such pyramids = $6 \times \frac{1}{3}$ area x ht

$$= 6 \times \frac{1}{3} \frac{\text{area of original triangle} \times \text{ht}}{4}$$

$$= \frac{\text{area of original triangle} \times \text{ht}}{2}$$

The interpretation is clever, though it is surprising that it was unknown to the able mathematicians of the Āryabhaṭa school. That Brahmagupta, Āryabhaṭa's critic, does not attack this verse and the verse supposedly embodying a wrong formula for the volume of a sphere lends some plausibility to the interpretation.

Barely a century later Brahmagupta knows the correct formula for the volume of a pyramid.

क्षेत्रफलं वेधगुणं समखातफलं हृतं त्रिभिः सूच्याः ।
मुखतलतुल्यभुजैक्यान्येकाग्रहृतानि समरज्जुः ॥

(*Br. Sp. Si.* XII. 44)

(The volume of a pit of uniform depth is the area (of section) multiplied by the depth. This divided by 3 is the volume of a sūcī, a figure tapering to a point. . . . .) The first part gives the formula for the volume of any (here regular only is meant) solid of uniform height. The second part gives the volume when the solid tapers uniformly to a point i. e., that of a pyramid or cone.

$$\text{The volume of a pyramid} = \frac{\text{Volume of a prism on the same base}}{3}$$

The method of derivation is not indicated.

The third part is elliptical and defies interpretation. Pṛthūdakasvāmin equates *mukhatalatulyabhujaikyam* to the products of the unequal breadths of the parallel strips of uniform depth into which he recommends a pit of non-uniform depth to be divided, and their depth. These products when divided by the sum of the widths of the strips and added together is the *samarajju*, the average depth. The interpretation gives us a correct method for finding the volume of a pit of non-uniform depth and hence with Pṛthūdaka we may say

सूत्राक्षराणां प्रायेण ईदृगेवार्थो यतः फलेन संवाद :

(This is the meaning of the words of the *sūtra*, since it agrees with the result.)

Mahāvīra's rule for calculating the volume of a pit of non-uniform depth is equally confused.

मुखतलयुतिदलमथ तत्संख्याप्तं स्यात् समीकरणम्

(*G. S. S.* VIII. 4)

(The sum of the (various) top dimensions with the (corresponding) bottom dimensions are halved. These are again added and divided by the number (of these halved quantities). This is the process of arriving at the average equivalent value.)

But the later writers make the method clear enough — that is, to take the average measurement.

8.4. *Frustum of a pyramid*

The formula for the volume of a frustum occurs for the first time in

मुखतलयुतिदलगुणितं वेधगुणं व्यावहारिकं गणितम् ।
मुखतल-गणितैक्यार्धं वेधगुणं स्याद् गणितमौत्रम् ॥
औत्रगणिताद्विशोध्य व्यवहारफलं भजेत्त्रिभिः शेषम् ।
लब्धं व्यवहारफले प्रक्षिप्य भवति फलं सूक्ष्मम् ॥

(*Br. Sp. Si.* XII 45, 46)

(The product of half the sums of the sides in the face and the base multiplied by the depth is the *vyāvahārika* (business measure) volume. Half the sum of the areas of the face and base multiplied by the depth is the *autra*[1] volume. The difference got by subtracting the *vyāvahārika* volume from the *autra* volume should be divided by three and the quotient should be added to the *vyāvahārika* volume. This will give the exact volume.)

The two rough values of the volume got by multiplying by the depth (1) the area of section got from the average lengths of the sides and (2) the average of the areas at the face and the base, are termed the *vyavahāra* volume and *autra* volume. If these are V_v and V_a,

the exact volume $= V_a + \dfrac{V_a - V_v}{3}$.

Now V_v (in a pyramidal frustum)

$= \left(\dfrac{a+a'}{2}\right)^2 \times h$ (where a and a′ are the sides of the base and the face and h is the height) and $V_v = \dfrac{a^2+a'^2}{2} \cdot h$

$$\therefore V = V_v + \frac{V_a - V_v}{3}$$

$$= \left(\frac{a+a'}{2}\right)^2 . h + \left\{\frac{a^2+a'^2}{2} - \left(\frac{a+a'}{2}\right)^2\right\} h$$

$$= h\left\{\frac{a^2+a'^2+2aa'}{4} + \frac{2a^2+2a'^2-a^2-a'^2-2aa'}{12}\right\}$$

[1]The purpose of introducing this *autra* volume seems to be mere ease of expression. The exact meaning of the word *autra* is still unknown.

$$= h \frac{3a^2+3a'^2+6aa'+a^2+a'^2-2aa'}{12}$$

$$= \frac{h}{3}(a^2+a'^2+aa')$$

And this is the correct formula for the volume of a frustum.

About Brahmagupta's formula for the volume of a pryamidal frustum J. L. Coolidge says "How did Brahmagupta ever reach the correct formula in this way . . . he may have heard the result somewhere. We have seen that it was known to the Egyptians. We shall presently see that the Chinese knew it also. He sets about justifying a traditional formula. The practical content is a prism standing on the mid-section. This is his first guess. A second guess would be a prism standing on the average of the two bases. These two give different values. Let us modify the first by adding to it a multiple of the difference between the two. He takes one third as a coeffecient perhaps because he thereby reaches the traditional result."[1] Let us hope Brahmagupta had a better basis.

The method can be used for finding the volume of any regular frustum. The formula is repeated by Mahāvīra.[2]

Śrīdhara gives an expression for the volume of the frus um of a cone.

$V = \frac{\pi}{24} \cdot h \{d_1^2+d_2^2+(d_1+d_2)^2\}$ where h is the height and d_1 and d_2 are the diameters of the base and the top. The conical frustum he has in view is a well.

मुखतलतद्योगानां वर्गैक्यकृते: पदं दशगुणाया:।
वेधगुणं चतुरन्वितविंशति-भवतं फलं कूपे॥

(*T. S.* 38)

(The square root of ten times the square of the sum of the squares of the diameter at the top, the diameter at the bottom and the sum of these diameters, when multiplied by the depth and divided by 24 gives the volume of a (circular) well.)

[1]*A History of Geometrical Methods* p. 16.

[2]But Mahāvira uses the word *aundra* instead of *autra*. The etymology of neither is clear

$$\text{i.e. } V = \frac{h}{24}\sqrt{10\,\{d_1^2+d_2^2+(d_1+d_2)^2\}^2}$$

$$= \frac{\pi \cdot h}{24}\left\{(d_1^2+d_2^2+(d_1+d_2)^2\right\}, \quad \text{since } \pi = \sqrt{10}$$

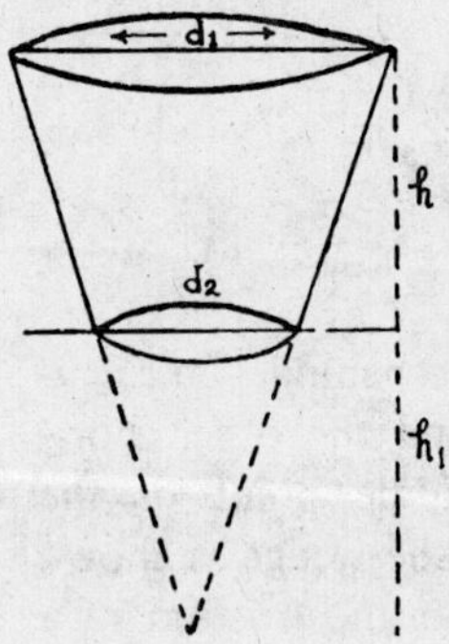

Fig. 1

This is, of course, based on the formula for the volume of a cone $\left(\text{Vol.} = \frac{\pi d^2 h}{4.3}\right)$ and a derivation similar to the derivation of the volume of a pyramid frustum is possible. The correct formula for the volume of a cone $\left(\frac{\pi d^2 h}{4.3}\right)$ is used in Śrīdhara's expression for the volume of a heap with circular base[1] calculated with $\pi = 3$.

8.5. *Vīrasena's method of infinite division for finding the volume of a cone-frustum*

Vīrasena (c. 710-790 A.D.)[2] details a method for finding the volume of the frustum of a cone in his *Dhavalā*.[3] If a and b are the diameters at the base and the top and h the height of the frustum, a cylindrical core of diameter b can be removed from the centre of the frustum. If the remaining sheath is slit open vertically, we get a wedge-shaped solid whose top edge is π. b in length and whose base is a trapezium with parallel sides equal to π . a and π . b and height $= \frac{a-b}{2}$.

Of its other two faces one is vertical and the other inclined.

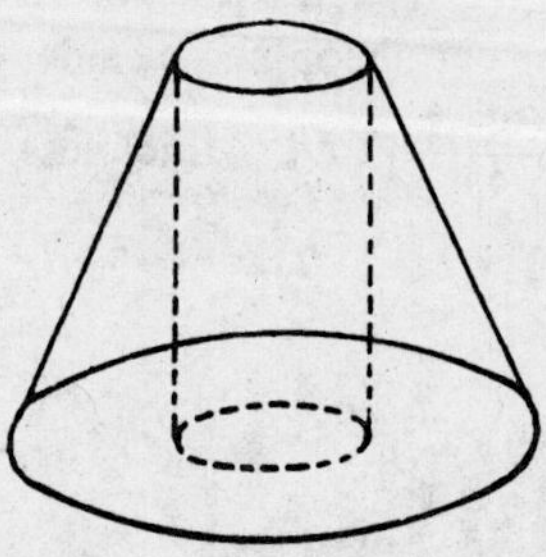

Fig. 2

[1] *T.S.* 61.

[2] *Jaina Sources of the History of Ancient India* by J.P. Jain. p. 189.

[3] *Ṣaṭkhaṇḍāgama* with the *Dhavalā* of Virasena. Ed. by Hiralal Jain. Jaina Sahitya Uddharaka Fund. Vol. IV. pp. 12-18.

A small wedge on a rectangular base of sides π b and $\frac{a-b}{2}$ can then be removed from the centre of this wedge.

The volume of this wedge = mean sectional area × height

$$= \text{length} \times \text{mean thickness} \times \text{height}$$

$$= \pi b \cdot \frac{\frac{2-b}{2} + o}{2} \cdot h$$

$$= \pi b \cdot \frac{a-b}{4} \cdot h$$

Two wedges on a triangular base are now left behind. To calculate their volumes, two planes, one vertical and the other horizontal are drawn passing through the middle point of the hypotenuse edge, which divide the section of the wedge as shown in the figure. Then the central part will be the frustum of a wedge with thickness at base $= \frac{a-b}{2}$,

height $= \frac{h}{2}$ and length of base $= \frac{\pi(a-b)}{2.2}$

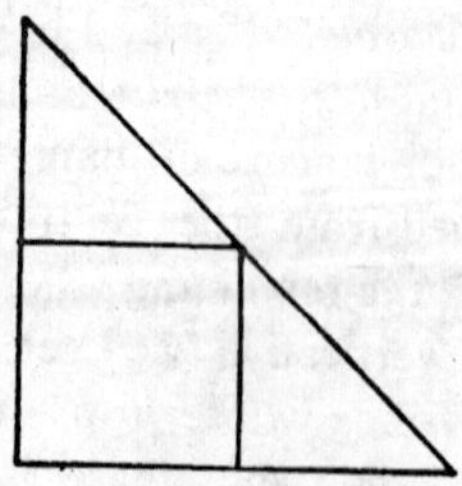

Fig. 3

If the two wedges so formed are placed one upon another with the thin top of one lying on the thick base of the other, we get a rectangular parallelopiped of sides $\frac{h}{2}$ and $\frac{\pi(a-b)}{4}$ and thickness $\frac{a-b}{2}$

Hence its volume $= \frac{\pi(a-b)}{4} \cdot \frac{a-b}{2} \cdot \frac{h}{2}$

$$= \frac{\pi(a-b)^2 \cdot h}{16}$$

The remaining 4 triangular wedges are again cut up similarly to yield 4 rectangular wedges which can be arranged in pairs to form parallelopipeds of sides $\frac{\pi(a-b)}{4}$, $\frac{h}{4}$ and $\frac{a-b}{4}$

The volume of these $=\frac{2 \cdot \pi (a-b)}{8} \cdot \frac{a-b}{4} \cdot \frac{h}{4} = \pi(a-b)^2 \cdot \frac{h}{64}$.

The remaining 8 triangular wedges when treated similarly produce 4 parallelopipeds of combined volume $\frac{\pi (a-b)^2 \cdot h}{64 \cdot 4}$

This process is to be repeated till the remainder triangular wedges are infinitely small and so can be neglected.

Then the total volume of all these parallelopipeds together

$$= \frac{\pi(a-b)^2}{16} \cdot h \left(1 + \frac{1}{4} + \frac{1}{4^2} \cdots \quad \cdots \right)$$

$$= \frac{\pi(a-b)^2}{16} \cdot h \cdot \frac{1 \cdot 4}{4-1}$$

$$= \frac{4}{3} \cdot \frac{\pi (a-b)^2 \cdot h}{16}$$

$\therefore$ The volume of the frustum

$$= \frac{\pi b^2}{4} \cdot h + \pi b \cdot \frac{a-b}{4} \cdot h + \frac{4}{3} \cdot \frac{\pi (a-b)^2}{16} h$$

$$= \frac{\pi . h}{4} \left(b^2 + ab - b^2 + \frac{a^2 + b^2 - 2ab}{3}\right)$$

$$= \frac{\pi . h}{4} \cdot \frac{3 b^2 + 3 ab - 3 b^2 + a^2 + b^2 - 2ab}{3}$$

$$= \frac{\pi . h}{4} \cdot \frac{a^2 - ab + b^2}{3}$$

Series mathematics has played an important part in the development of Indian mathematics, especially geometry and mensuration.

8. 6. Mahāvīra indicates a method for calculating the volume of a ditch round a circular, triangular or rectangular central space with the breadth of the ditch gradually decreasing.[1] The annulus is to be imagined to be stretched out into the frustum of a long rectangular wedge and then the volume to be calculated with the length and mean breadth and height.

[1] *G.S.S.* VIII. 19½—20½.

8.7. Mahāvīra arrives at the accurate volume of a tetrahedron via a curious formula for the approximate volume and a still more curious process of numerical manipulation.

भुजकृतिदलघनगुणदशपदनवहृत् व्यावहारिकं गणितम् ।
त्रिगुणं दशपदभक्तं श्रृंगाटकसूक्ष्मघनगणितम् ॥

(*G. S. S.* VIII 30½)

(The cube of half the square of the side is multiplied by 10. The root of the product divided by 9 gives the approximate volume. This multiplied by 3 and divided by $\sqrt{10}$ gives the exact volume.)

Hence the approximate volume $V_a = \dfrac{\sqrt{\left(\frac{a^2}{2}\right)^3 . 10}}{9}$ (where a is the edge of the tetrahedron)

$$= \frac{\sqrt{10} . a^3}{18\sqrt{2}}$$

Then the exact volume $= Vs . \dfrac{3}{\sqrt{10}}$

$$= \frac{\sqrt{10} . a^3}{18\sqrt{2}} . \frac{3}{\sqrt{10}}$$

$$= \frac{a^3}{6\sqrt{2}}$$

which is the correct expression.[1] But the source of the approximate formula is a baffling problem. Probably the product of half the sums of opposite sides was accepted as the approximate area of the base (i.e. $a . \frac{a}{2}$) and the approximate height was taken to be $\sqrt{a^2 - \left(\frac{2a}{3}\right)^2}$

$= \dfrac{\sqrt{5} . a}{3}$ when the volume will be $\dfrac{1}{3} a . \dfrac{a}{2} . \dfrac{\sqrt{5} . a}{3}$

$$= \frac{\sqrt{10} . a^3}{2.9.\sqrt{2}}$$

[1]Prof. Rangacharya thinks this is inaccurate because he equates the altitude of the tetrahedron with its edge.

With the correct expression for the volume of a pyramid known, the calculation of that of the tetrahedron with the altitude calculated as already shown is easy.

8.8. Besides these algebraical rules with numerical examples following, Mahāvīra sets a few numerical problems involving equality of two volumes. A sample is—

समचतुरस्रा वापी नवहस्तघना नगस्य तले ।
तच्छिखरात् जलधारा चतुरश्राङ्गुलसमानविष्कम्भा ॥
पतिताग्रे विच्छिन्ना तया घना सान्तरालजलपूर्णा ।
शैलोत्सेधं वाप्यां जलप्रमाणं च मे ब्रूहि ॥

(*G. S. S.* VIII. 35, 36)

(There is a tank at the foot of a mountain, square in shape and 9 *hastas* in each dimension. From its top a column of water whose section is a square of side one *angula* falls severed at the top (when the water touches the tank). By that column the tank is filled. Tell me the height of the mountain and the measure of water in the tank.)
The problem is repeated with the section of the water column changed into a circle, a triangle and a trapezium.[1]

8.9. Āryabhaṭa II, Nemicandra, Śrīpati, Bhāskara II and Nārāyaṇa have nothing new to add. But Śrīpati and Bhāskara give the volume of a frustum and a tapering solid in very general terms. Bhāskara's enunciation being the clearer is quoted.

मुखजतलजतद्युतिजक्षेत्रफलैक्यं हृतं षड्भिः ।
क्षेत्रफलं सममेव वेधहतं घनफलं स्पष्टम् ।
समखातफलत्र्यंशः सूचीखाते फलं भवति ॥ (*Lil.* 217)

(The sum of the areas got from the face elements, the base elements and the sums of these two, divided by six, is the area of the equivalent pit of uniform depth. This multiplied by the height is the exact volume. One third the volume of the pit of uniform depth is the volume of a *sūcī*.) This comprehends beside those for the prism and the cylinder the following formulae ;

1) The volume of the frustum of a pyramid on a rectangular base with base elements a and b and top elements a′ & b′ and height h

[2]*G.S.S.* VIII 37-4½.

$$= h\left\{\frac{a\,.\,b + a'\,.\,b' + (a + a')\,(b + b')}{6}\right\}$$

2) The volume of the frustum of a cone

$$= \frac{\left\{\frac{\pi\, d_1^2}{4} + \frac{\pi\,.\,d_2^2}{4} + \pi\frac{(d_1 + d_2)^2}{4}\right\}.\,h}{6}$$

$$= \pi\, h\,\frac{\{d_1^2 + d_2^2 + (d_1 + d_2)^2\}}{6\,.\,4}$$

(d_1 and d_2 are the base and top diameters.)

3) The volume of a pyramid on a rectangular base

$$= \frac{a\,.\,b\,.\,h}{3}$$

4) The volume of a cone $= \pi\frac{d^2\,.\,h}{4\,.\,3}$

8.10. *The Sphere*

Till we reach Āryabhaṭa I we do not come across any authentic mention of the sphere,[1] and Āryabhaṭa gives a wrong formula for the volume of a sphere.

समपरिणाहस्यार्धं विष्कम्भार्धहतमेव वृत्तफलम् ।
तन्निजमूलेन हतं घनगोलफलं निरवशेषम् ॥

(*A.B. Gaṇitapāda*. 7)

(Half the circumference multiplied by half the diameter is the area of a circle. That multiplied by its own root is the exact volume of a sphere.)

According to this the volume of a sphere $= \frac{\pi\, d^2}{4}.\sqrt{\frac{\pi\, d^2}{4}}$

$$= \frac{\pi^{3/2}\,.\,d^2}{8}$$

[1]The Jaina works refer to *Ghanaparimaṇḍala* but we do not know what exactly this means. Is it a cylinder or a sphere ? In the *Uttarādhyayana Sūtra* (c. 300 B.C.) there is a mention of a solid called *Iṣadprāgbhāra* "which resembles in form an open umbrella". Its thickness is greatest at the middle and decreases towards the margin till it is "thinner than the wing of a fly (*Uttarādhyayana Sūtra*, XXXVI, 59-60) The *Aupapattika Sūtra* says that the depth decreases at the rate of an *aṅgula* for every *yojana*. This suggests knowledge of the mensuration of a spherical segment. (*The Jaina School of Mathematics* by B. B. Datta. *Bull. Cal. Math. Soc.* vol. XXI, 1929).

Nīlakaṇṭha in his notes says that the formula is got by analogy. The square contents of a circle could be represented as a square whose side is the square root of the area. Hence the cubical contents of a sphere also should be capable of being represented as a cube on the same square base.[1] Elfering (see **8.3.** above) interprets this verse also differently so as to give the correct expression for the surface area (not volume) of a hemisphere.

Brahmagupta does not deal with the sphere. Śrīdhara's formula for the volume of a sphere is

गोलव्यासघनार्धं स्वाष्टादशभागसंयुतं गणितम् ।

(*T.S.* 56)

(Half the cube of the diameter of a sphere plus the eighteenth part of itself is its volume.)

$$\text{i.e. Volume} = \frac{d^3}{2} + \frac{d^3}{2 \cdot 18} = \frac{19}{36} d^3$$

This is obtainable from the correct expression for the volume of a sphere by substituting

$$\pi = \sqrt{10} = \frac{19}{6} \left(\text{For } \sqrt{10} = 3 + \frac{1}{6} = \frac{19}{6} \right)$$

$$\text{Then volume} = \frac{\text{Surface area} \times \text{diameter}}{6}$$

$$= \frac{\pi \cdot d^2 \cdot d}{6} = \frac{\sqrt{10} \cdot d^3}{6} = \frac{19\ d^3}{6 \cdot 6} = \frac{19\ d^3}{36}$$

Mahāvīra sets down

व्यासार्धघनार्धगुणा नव गोलव्यावहारिकं फलम् ।
तद्दशमांशं नवगुणमशेषसूक्ष्मं फलं भवति ॥

(*G. S. S.* VIII $28\frac{1}{2}$)

(Nine multiplied by half the cube of the radius is the working formula for the volume of a sphere. Its tenth part multiplied by 9 is the exact volume.)

$$\text{i.e. } V_a = r^3 \cdot \frac{9}{2} = \frac{3}{2} \cdot \pi r^3,$$

(since $\pi = 3$ for practical purposes according to Mahâvīra)

[1] अस्मिन् फले मूलिते पुनस्तन्निर्मितचतुरश्रबाहुः स्यात् । एवं वृत्तक्षेत्रेण समचतुरश्रं सम्पादनीयम् । एवं घनगोलस्य समद्वादशाश्रत्वमापन्नस्यापि तच्चतुरश्रत्वमापन्नस्यापि तच्चतुरश्रबाहुतुल्य ऐव द्वादश बाहवः ॥

and $V_e = r^3 \cdot \frac{9}{2} \cdot \frac{9}{10} = \frac{81r^3}{20}$

Since the accurate value of π is $\sqrt{10}$ this reduces to

$$V_e = 1{\cdot}3\ \pi\ r^3 \text{ nearly.}$$

This is not far removed from the correct formula $V = \frac{4}{3} \cdot \pi\ r^3$.

Āryabhaṭa II's formula is

कन्दुकपिण्डस्य घनो दलितः स्वाष्टादशांशसंयुक्तः ।
धनहस्ता., ...

(*Ma. Si.* XIV. 108)

(Half the cube of the thickness of a ball combined with its own eighteenth part is its volume)

i.e. Volume of a sphere $= \frac{d^3}{2} + \frac{d^3}{2.18}$

$$= \frac{19\ d^3}{36} = \frac{38}{9} r^3$$

$$= 1.\ 34.\ \pi.\ r^3$$

Bhāskara and the later mathematicians give correct formulae for both the surface area and the volume of the sphere.

वृत्तक्षेत्रे परिधिगुणितव्यासपादः फलं य-
त्क्षुण्णं वेदैरुपरिपरितः कन्दुकस्येव जालम् ।
गोलस्यैवं तदपि च फलं पृष्ठजं व्यासनिघ्नम्
षड्भिर्भक्तं भवति नियतं गोलगर्भे घनाख्यम् ॥

(*Līl.* 201)

(In a circle, the circumference multiplied by one-fourth the diameter is the area, which, multiplied by 4, is its surface area going round it like a net round a ball. This (surface area) multiplied by the diameter and divided by 6 is the volume inside the sphere.)

i.e. area of a circle = circumference . $\frac{d}{4}$

$$= \frac{\pi\ d^2}{4} = \pi\ r^2$$

The surface area of a sphere = 4. area of its great circle

$$= 4\,\pi\, r^2$$

Volume of a sphere = Surface area $. \frac{2r}{6}$

$$= \frac{4}{3}\pi\, r^3$$

Bhāskara explains the mode of derivation of the formula for the surface area in the Golādhyāya of his *Siddhānta-Śiromaṇi* (*Vāsanā* under V. 54-57).

"Take a sphere of clay or wood to represent the earth. Let its circumference be 21600 *kalas*. With the point at the top as centre and $\frac{1}{96}$th of the circumference i.e. 225 *kalas* as the radius draw a circle on the surface of the sphere. Similarly draw circles with 2. 225, 3. 225 . . . 24.225 *kalas*. These circles will cover half the sphere, dividing the surface into 23 strips and a small circle at the top. The radii of these circles in their own planes will obviously be the sine-chords of the arcs of the great circle of the sphere. These are already known. If the strips going round the sphere are cut and spread, they will be trapezia with the parallel sides equal to the circumferences of the consecutive circles and 225 *kalas* as the altitude. The sum of the areas of these trapezia and that of the circle at the top will be the surface area of half the sphere."

The same method is also given in the text of the *Golādhyāya* three or four verses later, with the addition of longitudinal lines cutting the surface into lunes.

गोलस्य परिधिः कल्प्यो वेदघ्नज्यामितेर्मितः ।
मुखबुध्नगरेखाभिर्यद्वदामलके स्थिताः ॥
दृश्यन्ते वप्रकास्तद्वत् प्राग्‌क्तपरिधेर्मितान् ।
ऊर्ध्वाधःकृतरेखाभिः गोले वप्रान् प्रकल्पयेत् ॥
तत्रैकवप्रकक्षेत्रफलं खण्डैः प्रसाध्यते ।
सर्वज्यैक्यं त्रिभज्यार्धहीनं त्रिज्यार्धभाजितम् ।
एवं वप्रफलं तत् स्याद् गोलव्याससमं यतः ।
परिधिव्यासघातोऽतो गोलपृष्ठफलं स्मृतम् ॥

Siddhānta Śiromani, Golādhyāya, Bhuvanakośa 58—61)

(The circumference of the sphere is to have as many units in it as 4 times the number of sine-chords in the table of sines (i.e. 24.4 = 96). With lines starting from the top and ending at the bottom, the surface should be divided into as many parts as there are units in the circumference—into parts which resemble the natural divisions on the surface of an *āmalaka* (myrobalan) fruit. Then the area of one such part (*vapraka*) should be found out by dividing it into (n) parts again (by horizontal lines at distances of one unit from each other). This area of a *vapraka* will be equal to the sum of all the tabular sine-chords diminished and divided by half the radius i.e. equal to the diameter. Hence the surface area is the product of the circumference and the diameter.

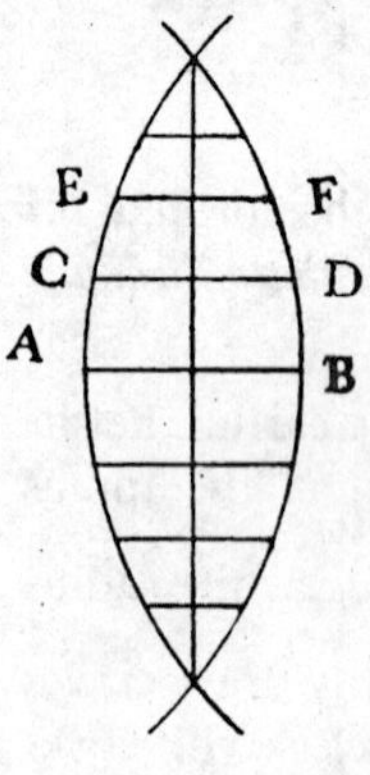

Fig. 4

The area of one *vapraka* = the sum of the areas of the trapezia like A B C D, C D E F

$$=2\left\{\frac{AB+CD}{2}.\text{altitude}+\frac{CD+EF}{2}.\text{altitude}+\dots\dots\right\}$$

$$=2\left\{\frac{\frac{2\pi r.\sin na}{2\pi r}+2\pi r.\frac{\sin(n-1)a}{2\pi r.}}{2}.\text{alt.}+\right.$$

$$\frac{2\pi r.\sin(n-1)a+2\pi r.\sin(n-2)a,}{2.2\pi r}.\text{alt.}+\dots\dots\dots\dots\dots$$

$$\left.+\frac{2\pi r.\sin a}{2.2\pi r}\text{alt.}\right\}$$

$$=2.\frac{2\Sigma r.\sin na-\sin na}{2r}.1$$

$$=\frac{(\Sigma r.\sin na-\frac{r}{2})}{r/2}$$

This has been equated by Bhāskara to the diameter[1] of the

[1]Bhāskara's knowledge of the mode of derivation is uncertain. See also *Infinitesimal Calculus in Indian Mathematics—its origin and development* by P.C. Sen Gupta—*Jou. Dept. Let.* XXII. 1932 p. 1 ff.

sphere and verified with respect to the sphere of radius 3438 by actual calculation.

i.e. The area of one *vapraka* = d

$\therefore$ The whole surface area = d . 2 π . r

$$= 4 \pi r^2$$

The rationale for the expression for the volume is given by Bhāskara in the 'Vāsanā' under these same verses.

"पृष्ठफलसंख्याकानि रूपबाहूनि व्यासार्द्धतुल्यवेधानि सूचीखातानि गोलपृष्ठे प्रकल्प्यानि सूच्यग्राणां गोलगर्भे संपातः । एवं सूचीफलानां योगो घनफलमित्युपपन्नम् ।"

(As many pyramids as there are units in the surface-area with bases of unit side and altitude equal to the semi-diameter should be imagined on the surface of the sphere. Then it is proper that the sum of the volumes of the pyramids should be the volume of the sphere.)

The volume of one such pyramid $= \dfrac{1^2 . r}{3}$

The volume of all these together $= \dfrac{4 \pi r^2 . r}{3}$

$$= \frac{4 \pi r^3}{3}$$

In Europe it was Kepler (early 17th Century) who found out this method of derivation of the expression (Carl. B. Boyer—*The History of the Calculus*. p. 108)

According to Nārāyaṇa the surface area of a sphere is thrice the square of the diameter, (*G. K. Khātavyavahāra*. 6)

$$\text{that is } = 3d^2 = 4.3.r^2$$

$$\text{and the volume} = \text{surface area} \times \frac{\text{diameter}}{6}$$

$$= \frac{4.3.r^3}{3}$$

Evidently he has not bothered to give the *sthūla* and *sūkṣma* values separately for the surface area and volume of a sphere (as he does for the areas of plane figures) and is satisfied with the gross value obtained with $\pi = 3$.

8.11. *Surface area of a sphere by integration*

The *Yuktibhāṣā* derives the expressions for the surface area and volume of a sphere with the help of the differentiation-integration, which the Āryabhaṭa school uses with such success on other occasions.

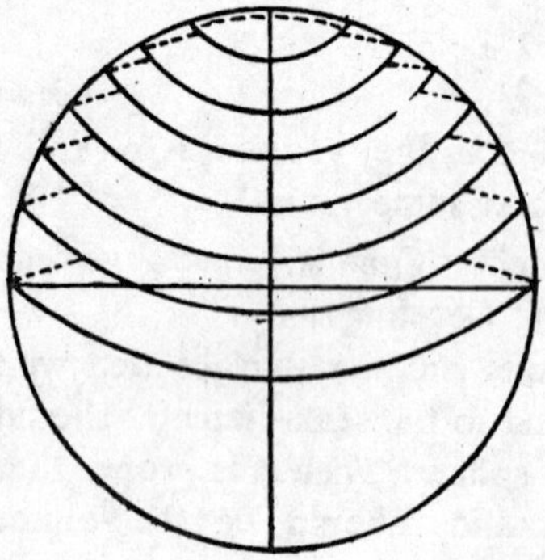

Fig. 5

[1]The surface of the hemisphere is to be divided into a large number of circular strips by drawing circles parallel to the horizontal great circle of the sphere at equal distances. Then as in Bhāskara's method, if these strips are cut open and spread, these will be trapezia with parallel sides equal to the circumferences of the consecutive circles on the surface of the sphere, and altitude = the chord of the arc between two consecutive circles.

The area of any such trapezium =

$$\frac{\text{circumference of upper circle} + \text{cirumference of lower circle.}}{2}\text{chord}$$

$$= \text{circumference of the circle through the middle of the strip} \times \text{chord.}$$

$\therefore$ The surface area of the hemisphere = the sum of the circumferences of the middle circles $\times$ chord.

Now the radii of these circles will be the sine-chords in a quarter of the great circle of the sphere. Hence if C is the circumference of the great circle of the sphere, and r the radius, the circumferences of the circles through the middles of the strips will be the $\frac{\text{corresponding sine-chord. C}}{\text{r}}$. Hence the sum of the circumferences

$$= \frac{\text{C}}{\text{r}} \quad \text{the sum of sine-chords}$$

As has already been shown

[1] *Y.B.* pp. 272-282.

$\dfrac{\text{Ist sine chord} \times (\text{whole chord of the Ist arc})^2}{r^2}$ = the difference between Ist and 2nd sine differences.

$\dfrac{\text{2nd sine chord} \times (\text{whole chord of the Ist arc})^2}{r^2}$

= the difference between the 2nd and 3rd sine differences

... ...

... ...

$\dfrac{(\text{Whole chord of the Ist arc})^2}{r^2}$. sum of the sine-chords

= sum of the differences of the sine differences

= sum of the 2nd differences of the sine chords.

= the difference between the Ist and 2nd sine chords minus the difference between the last and penultimate sines.

If the arcs between the circles are very small the latter quantity will be nearly zero.

Hence the sum of the sine-chords

= The difference between the Ist and 2nd sine chord

$\dfrac{r^2}{\text{whole chord}^2}$.

$= \dfrac{\text{whole chord} \times r^2}{(\text{whole chord})^2}$ since for small arcs, the sine-difference is almost equal to the arc difference.

Hence the surface area of the hemisphere

$= \dfrac{C}{r}$. sum of the sine chords . chord

$= \dfrac{C}{r} . \dfrac{r^2}{\text{chord}}$. chord .

$= C . r$

Then the surface area of the whole sphere $= 2\ C . r = 2 . 2\pi\ r . r$

$= 4\pi r^2$

8.12. *Volume of a sphere by integration*

For finding the volume,[1] the sphere is to be cut into circular laminae by planes passing through circles drawn on the surface of the sphere, as for finding the surface area. These are to be of the same and uniform thickness equal to one unit. Then the volume of any lamina

= the area of the circle through its middle . thickness

$= \pi s^2 . 1$ where s is the radius of the circle and is = the corresponding sine chord .

Hence the volume of the whole sphere

= sum of the areas of the laminae . 1

$= \pi (s_1^2 + s_2^2 + \ldots\ldots + s_n^2) . 1$ where $s_1, s_2 \ldots\ldots$ are the sine chords beginning from one pole to the other of the sphere.

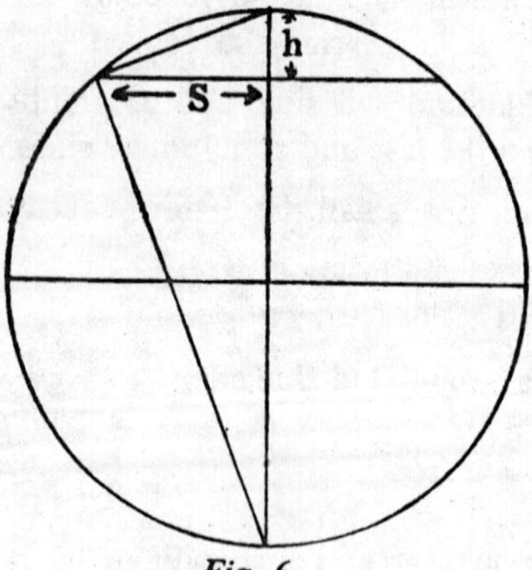

Fig. 6

Now if h is the height of double the arc whose sine-chord is s and d is the diameter of the circle

$$s^2 = h(d-h)$$

$$= \frac{2h(d-h)}{2}$$

$$= \frac{(h+d-h)^2 + \{h^2+(d-h)^2\}}{2}$$

$$= \frac{d^2 - \{h^2+(d-h)^2\}}{2}$$

Hence $s_1^2 + s_2^2 + \ldots\ldots s_n^2$

$$= \frac{d^2 - \{h_1^2+(d-h_1)^2\}}{2} + \frac{d^2 - \{h_2^2+(d-h_2)^2\}}{2} + \ldots\ldots + \frac{d^2 - \{h_n^2+(d-h_n)^2\}}{2}$$

$$= \frac{n}{2} . d^2 - \frac{\Sigma h_n^2}{2} - \frac{\Sigma(d-h_n)^2}{2}$$

Now, if the sphere is divided into an infinitely large number of laminae so that the thickness (one unit) of each lamina is infinitely small.

[1] *Y.B.* pp. 282-290.

$h_1 = \triangle d,\ h_2 = 2 \triangle d,\ h_3 = \triangle d \ldots\ldots\ldots\ldots$
and $d - h_1 = h_{n-1},\ d - h_2 = h_{n-2} \ldots\ldots\ldots\ldots,\ d - h_{n-1} = h_1$

i.e. $\quad \Sigma\, h_n = \Sigma\, (d - h_n)$

Again $\Sigma\, h^2_n = (\triangle\, d)^2 + (2 \triangle\, d)^2 + \ldots\ldots + (n \triangle d)^2$

$$= \frac{d^3}{3} \text{ (n being equal to d)}$$

$$\therefore\ \Sigma s^2_n = \frac{n}{2} \cdot d^2 - \frac{2\, d^3/3}{2} = \frac{d^3}{2} - \frac{d^3}{3} = \frac{d}{6}$$

Hence volume of the sphere $= \pi\, \Sigma s^2_n \cdot 1$

$$= \pi\, \frac{d^3}{6} = \frac{4\, \pi\, r^3}{3}$$

As the editors of the *Yuktibhāṣā* show, these methods of finding the surface area and volume of a sphere are the same as the modern integration methods which were discovered in Europe by Newton and Leibnitz.

CHAPTER IX

GEOMETRICAL ALGEBRA

9.1. The practice of representing and solving algebraic and arithmetical problems geometrically is as old as geometry itself. The *Śulbasūtras* find $\sqrt{2}$, $\sqrt{3}$ etc. (theoretically the square root of any number) with the help of squares and rectangles. For $\sqrt{2}$, the *Śulbasūtras* have

समस्य द्विकरणी

(*Āp. Sl.* I. 5)

(The diagonal of a square of unit side will be $\sqrt{2}$.) The Sanskrit word for $\sqrt{2}$ is *dvikaraṇī*, the maker of twice the area.

प्रमाणं तिर्यक् द्विकरण्यायाम: तस्याक्ष्णयारज्जुस्त्रिकरणी । (*Āp. Sl.* II. 2)

(One measure is the horizontal side and the *dvikaraṇī* is the length. Its diagonal chord will be the *trikaraṇī*, the maker of an area of 3 measures) i.e., $\sqrt{3}$. In this way by drawing rectangles with sui able sides the square root of any number can be found. The method is extended to fractions also. For finding $\sqrt{\frac{1}{3}}$ i.e. the one-third maker, Āpastamba supplements the foregoing *sūtra* by

तृतीयकरण्येतेन व्याख्याता, प्रमाणविभागस्तु नवधा । (*Āp. Sl.* II. 3)

(The one third-maker has been explained by this. The division is in nine parts.) The meaning of this cryptic statement is made clear by the corresponding *sūtras* in the *Kātyāyana Śūlbasūtra*.

तृतीयकरण्येतेन व्याख्याता, प्रमाणविभागस्तु नवधा । करणीतृयीयं नवभाग: नवभागास्त्रयस्तृतीयकरणी

(*K. Sl.* II. 15-18)

(The one-third maker is expounded by this. The division of the measure (of the area) is into 9 parts. One-third of the *karaṇī* i.e. the side of the square makes one-ninth (of the area). Three ninth parts have one-third as its *karaṇī* or maker).

The square drawn on one as side is to be divided into 9 equal parts. 3 of these small squares are to be combined into one square. The side of this square will be $\sqrt{\frac{1}{3}}$.

The same method can be and is used for evaluating $\sqrt{a^2+b^2}$ and $\sqrt{a^2-b^2}$ where a and b are any rational numbers. A rectangle is drawn with a and b as sides. Then its diagonal will be $\sqrt{a^2+b^2}$. Here (as also in evaluating $\sqrt{a^2-b^2}$), the purpose of the Śulbasūtras is really more geometrical i.e. to combine two squares into an equivalent square and the method is

ह्रसीयसः करण्या वर्षीयसो वृध्रमुल्लिखेत् वृध्रस्याक्ष्णयारज्जुरुभे समस्यति ।

(*Āp. Sl.* II. 4)

(A segment should be cut off from the bigger square by the side of the smaller square. The diagonal chord of the segment will combine the two squares).

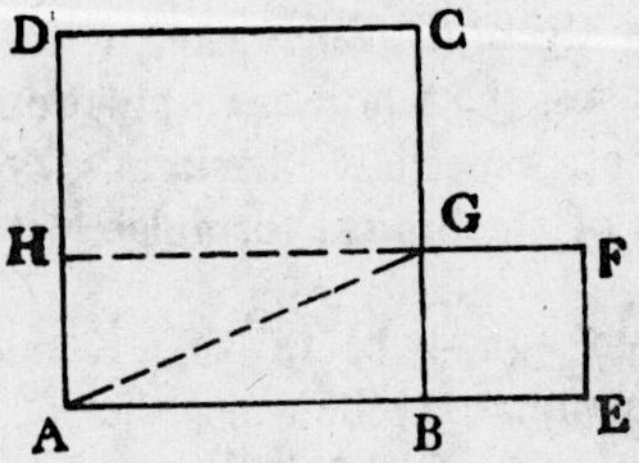

Fig. 1

To get a square equal to the difference of two squares, the construction[1] is intially the same as that for finding the sum of two squares. But instead of joining the diagonal, the lower side AB of the rectangular segment cut off, i.e. ABFH, is to be raised diagonally with one end (A) fixed, till the other end touches the upper side H F of the segment at B′. Then HB′ is the side of the required square.

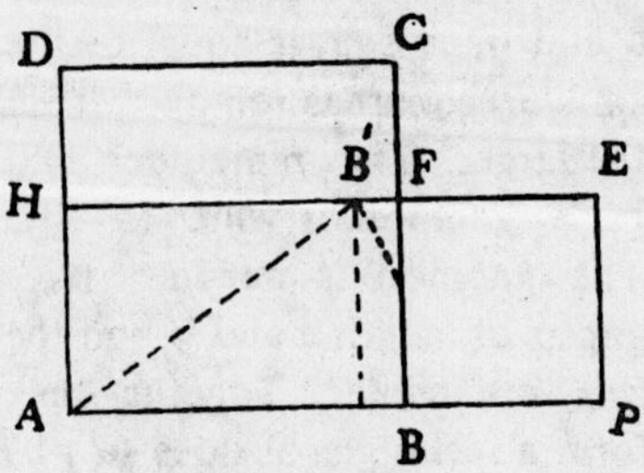

Fig. 2

For $HB'^2 = AB'^2 - AH^2$.

The construction is ingenious though basically it is a simple application of the theorem of the square of the hypotenuse.

[1]*Āp. Sl.* II. 5.

9.2. In later days, most probably in the wake of the tradition of the *Śulbasūtras*, geometry was called in to validate algebraical rules, but there is a shift in the emphasis. The *Śulbasūtra* writers were primarily interested in geometrical constructions, the implied algebraical truths coming in by the side door. But with the later mathematicians the algebraical results are the most important, the geometrical figures being merely an aid to make the algebraical results the more immediately convincing, or to prove the results.

In Brahmagupta, Mahāvīra and Bhāskara II such use of geometry is limited. The section on rational triangles and quadrilaterals, to which elaborate attention is given by all these authors, may be called geometrical algebra. But, except for the drawing of the figures themselves, the whole manipulation is algebraical. Later, the commentators especially Bhāskara's, do give diagrammatical corroborations of algebraical formulae.

9.3. Thus to demonstrate $a^2-b^2=(a+b)(a-b)$, Kṛṣṇa Daivajña[1] (1606 A.D), the author of the *Navāṅkura* commentary on the *Bījagaṇita* of Bhāskara takes two squares of sides a and b. A square equal to the smaller square is removed from the larger. The remainder will be a gnomon of width (a—b). The gnomon is cut into rectangles of length a and b and the two are joined. Together they form a rectangle of sides (a+b) and (a—b).
Hence $a^2-b^2=(a+b)(a-b)$

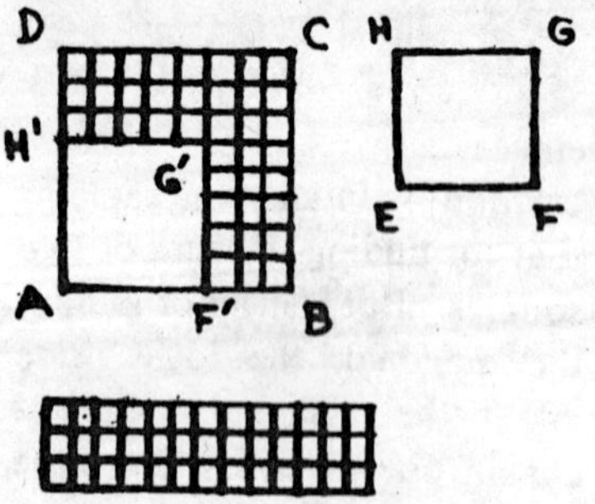

Fig. 3

This method of demonstration is adopted by Gaṇeśa also in his commentary (1545 A.D.) on the *Līlāvatī*.

[1]*Bhāskarīya Bījagaṇitam*. Anandasrama Seriies No. 99. p. 150.

The rationale for $(a+b)^2-a^2-b^2=2ab$ is also given diagramatically.[1] From the square of side $a+b$ when squares with sides a and b are removed, two rectangles of sides a and b remain.

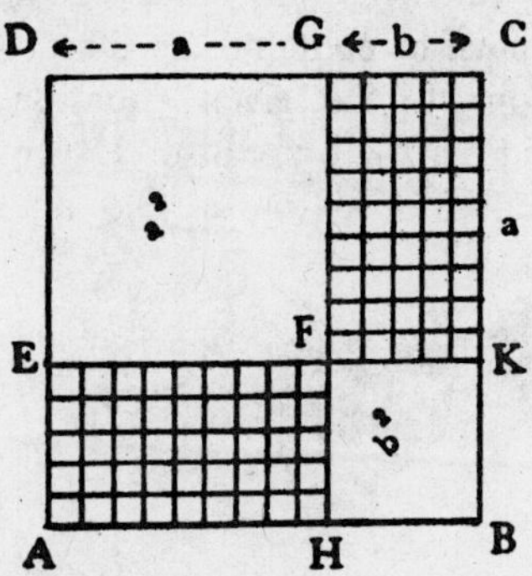

Hence $(a+b)^2-a^2-b^2=2ab$

Fig. 4

For indeterminate equations of the form $ar+by+c=xy$, Kṛṣṇa gives a geometrical method, saying

अत्रोपपत्तिराचार्यैर्लिखिता अस्ति । किन्तु लेखकादिदोषादुपदेशविच्छित्त्या च संप्रति सा न स्वकार्यक्षमा । अत इयं भावितोपपत्तिर्विविच्योच्यते[2]

(Here the rationale has been written by the master. But due to the errors of the scribes and a break in the tradition of instruction, that (rationale) is not now capable of fulfilling its purpose. Hence the rationale for *Bhāvita* is given in detail.)

The product x y may be represented by a rectangle of sides x and y. This should contain a xs b ys and the absolute number of unit areas, c. Mark off a xs in the figure. Now the remaining rectangle being only y—a in length, the full y's cannot be marked off. Hence mark off (y—a) b times. Here each y—a is less then y by 'a' units

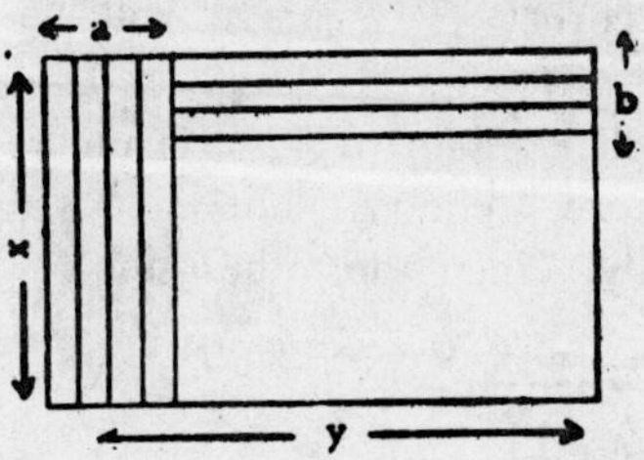

Fig. 5

$\therefore$ b (y—a)' s are less than b y's

[1]Ibid. p. 152.

[2]Ibid. p. 196. E.T. Bell who says (*Development of Mathematics* 1945 p. 125) that the letter *Bhā* is the sign of multiplication in Indian Mathematics is perhaps misled by the term *Bhā* appearing with the product x.y in the context of *Bhāvita*.

by ab units. If on the other hand, b full y's are first marked off, we will be able to mark off (x—b)'s only and again the deficiency is ab. Hence, this ab must be contained in the remaining part of the diagram. According to the given equation the rectangle x y should contain besides a x & b y, c also. Hence the rest of the diagram must be composed of ab and c. Let one side of this remainder rectangle be k.

Then the other side is $\frac{ab+c}{k}$

Then $k + a = y$

$$\frac{a\,b + c}{k} + b = x$$

If a and b are negative, the method has to be changed slightly. Since $-ax - by + c = xy$, add $ax + by$ to both sides.

Then $xy + ax + by = c$.

The diagrammatic representation of the left-hand-side will be a rectangle xy with a x's and b y's annexed to it. To fill up the empty corner a . b units will be required. Hence the bigger rectangle will contain c + ab units. As before, choose one side k arbitrarily. Then the other side

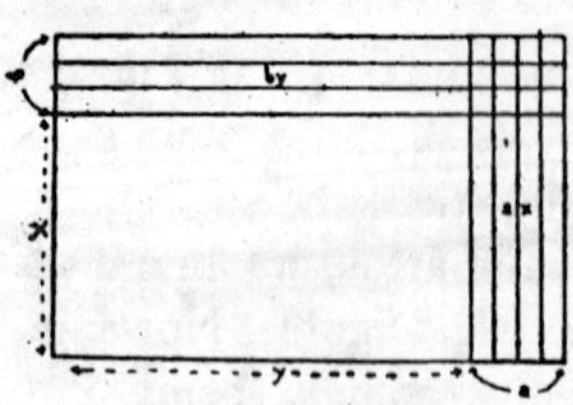

Fig. 6

$$= \frac{ab + c}{k}$$

Hence $x = k - b$

and $y = \frac{a\,b + c}{k} - a$

When c is negative a and b may be greater then y and x. In this case the two rectangles ax and b y are to be superposed at one corner.

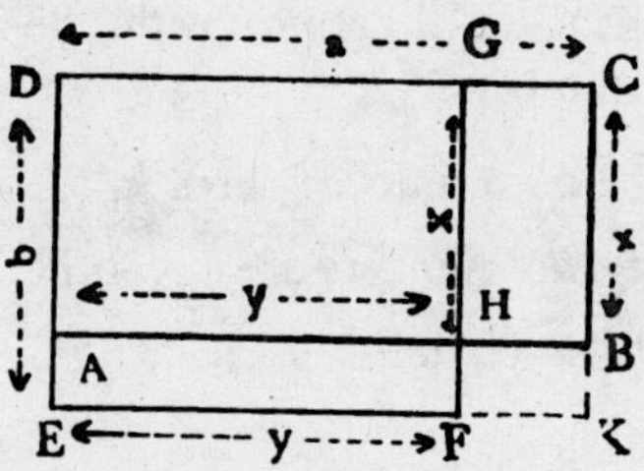

Fig. 7

Then one x y is to be added to this figure to get a x + b y and since a x + b y — c = xy, the figure should be equal to c. Now, if the empty corner is filled up, the area of the whole rectangle = ab.

∴ The area of the rectangle H K = ab — c.

So, if k is one side of the rectangle, the other side is

$$\frac{a\,b - c}{k}$$

Then b — k = x and $a - \frac{a\,b - c}{k} = y$.

9.4. In the Āryabhaṭa school of mathematics the use of geometry to substantiate algebraical results is a regular feature.

Bhāskara I offers a geometrical explanation for $\frac{1}{4} \cdot \frac{1}{5} = \frac{1}{20}$.

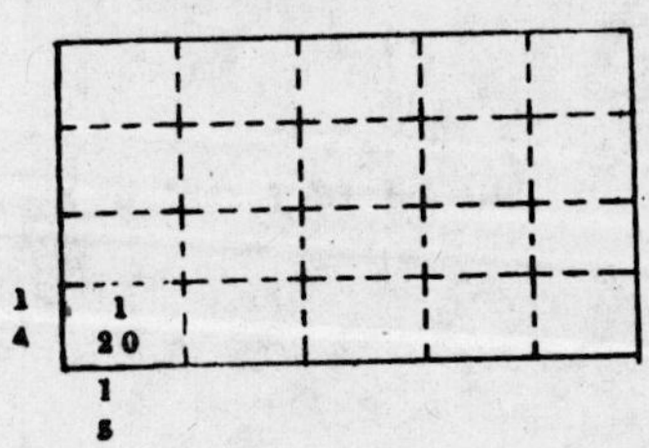

Fig. 8

[1] If one side of a rectangle is divided into 5 equal parts and the other into 4 and parallels are drawn through these points of division, the whole figure is divided into 20 parts and the sides of each small rectangle are $\frac{1}{5}$ and $\frac{1}{4}$ of the corresponding sides of the original rectangle.

$$\therefore \frac{1}{5} \cdot \frac{1}{4} = \frac{1}{20}.$$

9.5. The full bloom of this geometrico-algebraical imagination found in Nīlakaṇṭha Somayājin and his followers, the authors of the *Kriyākramakarī* and the *Yuktibhāṣā*.

[1] *Āryabhaṭīyabhāṣyam* p. 55. Also *Y.B.* pp. 37-38.

The product of any two unequal numbers can be represented as a rectangle (*ghātakṣetra*= multiplication diagram) with the sides respectively equal to the given numbers.[1]

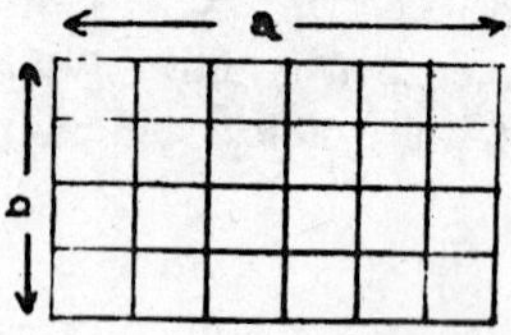

There will be a . b units of area in the figure. If the two numbers are equal the diagram will be a square.

Fig 9

For a $(b + b_1)$ the representation will be as in fig. 10, i.e., the strip A′ B containing b_1 rows of 'a' small squares is to be tacked on to the rectangle A C representing a . b. Similarly for a $(b-b_1)$, b_1 rows each containing 'a' small squares have to be removed from the rectangle.[2]

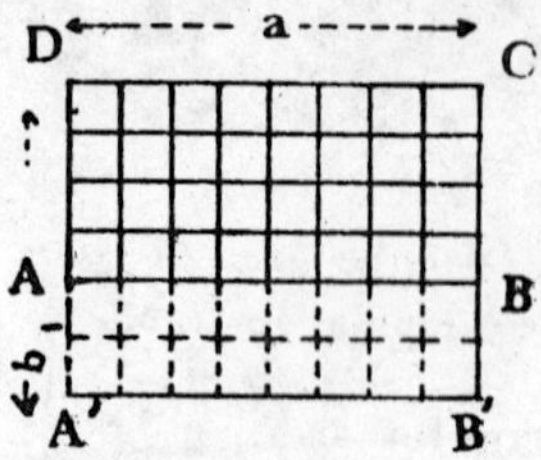

Fig. 10

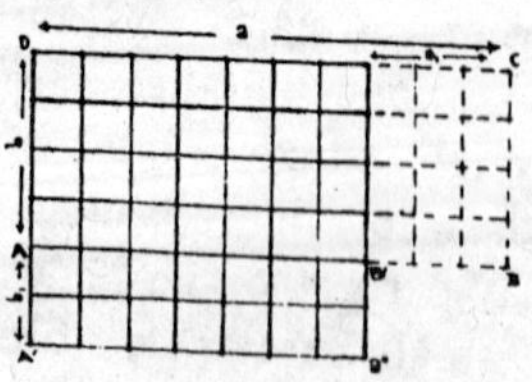

$(a-a_1)(b+b_1)$ will be represented[3] by fig. 11.

Fig. 11

Another multiplication result demonstrated geometrically[4] is

$$ab = (a + a_1)\left(b - b \cdot \frac{a_1}{a + a_1}\right)$$

This is shown with $a = 12$, $a_1 = 4$ & $b = 20$

[1] *Y.B.* pp. 7 & 8.
[2] Ibid p. 9.
[3] Ibid p. 10.
[4] Ibid pp. 12-13.

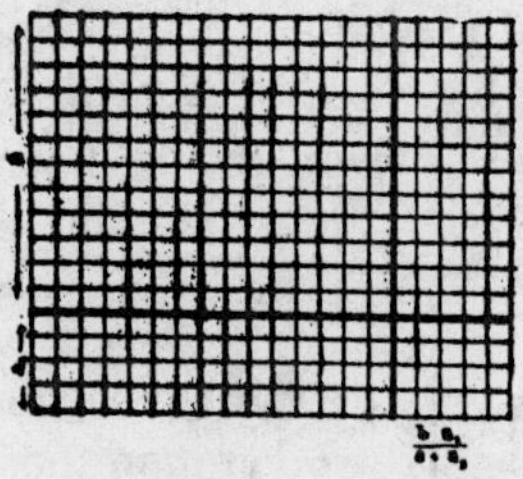

Fig. 12

9.5a. Division, being the inverse of multiplication, is also capable of being demonstrated diagrammatically.[1] The dividend can be represented as a rectangle with one side = the divisor. Hence, if we go on placing rows of squares equal in number to the divisor, each parallel to and touching the previous one, till the total number of small squares is equal to the dividend, the number of the rows will be the quotient.

9.6. Next is given the familiar demonstration for

$(a + b)^2 = a^2 + b^2 + 2\,ab.$

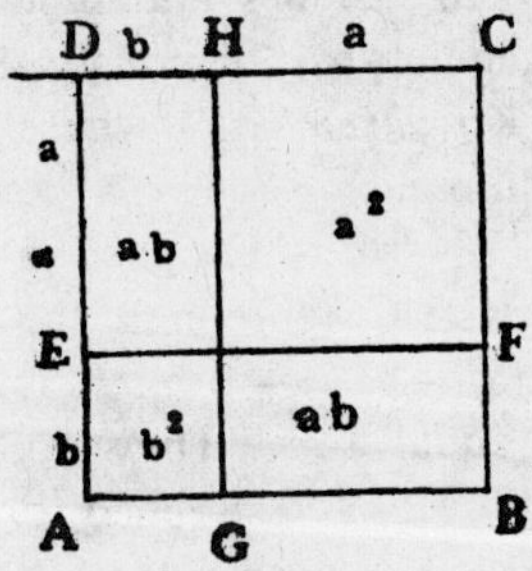

Fig. 13

ABCD represents $(a+b)^2$. Through points of division of the sides parallels are drawn, when the squares a^2 and b^2 and 2 rectangles result. The sides of the rectangles being a and b, the result $(a + b)^2 = a^2 + b^2 + 2\,ab$ is established. The method can easily be extended to the squares of the sums of any number of terms.

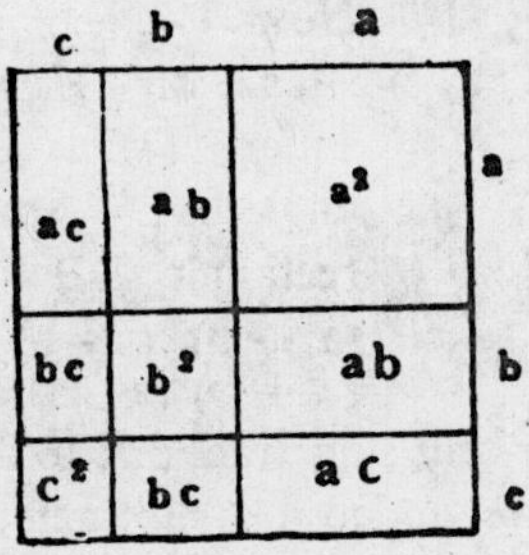

Fig. 14

[1] Ibid p. 15

Similarly fig. 14 can easily establish the equation $(a+b+c)^2 = a^2+b^2+c^2+2ab+2bc+2ac$.

For showing that $4\ ab + (a-b)^2 = (a+b)^2$ the four rectangles a . b are arranged as shown in the figure.[1] Then the side of the square at the middle is evidently $(a-b)$.

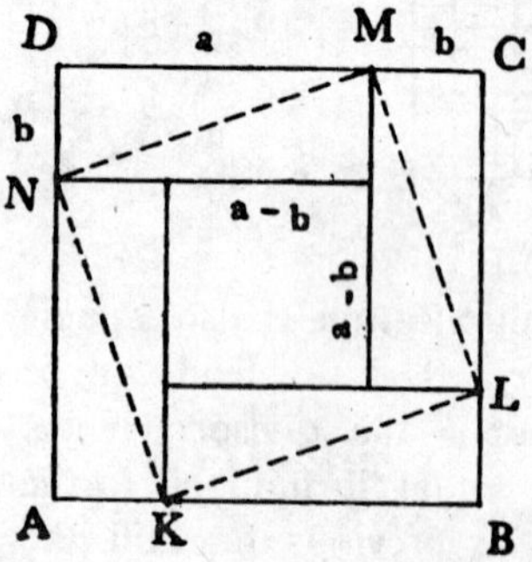

Fig. 15

If the diagonals of these rectangles other than those which pass through the corners of the square are joined, the same figure will serve to substantiate another result. Since the diagonals divide each rectangle into equal halves the inner square KLMN i.e. the square on the diagonal $= (a+b)^2 - 2\ ab$

$= a^2 + b^2 + 2\ ab - 2\ ab$

$= a^2 + b^2$

(The author of the *Yuktibhāṣā* says (p. 35) that cubes and cube roots are not treated in his work. The geometrical demonstrations for $(a+b)^3$ etc. are given in the *Kriyākramakarī* whose author is perhaps a pupil of Nīlakaṇṭha himself.)

9.7. Next the formula $(a+b)(a-b) = a^2 - b^2$ is assigned a a geometrical explanation.[2] The rectangle A C represents the *ghātakṣetra* $(a+b)(a-b)$. DK is marked off equal to a. The strip KGBC is removed and applied to the same figure along AG to occupy the new position AELM. Now $DE = a-b+b = a$. The square on DK is completed. Then square DF = rectangle DG + rectangle AL + square MF. i. e. $a^2 =$ rectangle DB + Sq. MF i.e. $a^2 = (a+b)(a-b) + b^2$ or $a^2 - b^2 = (a+b)(a-b)$

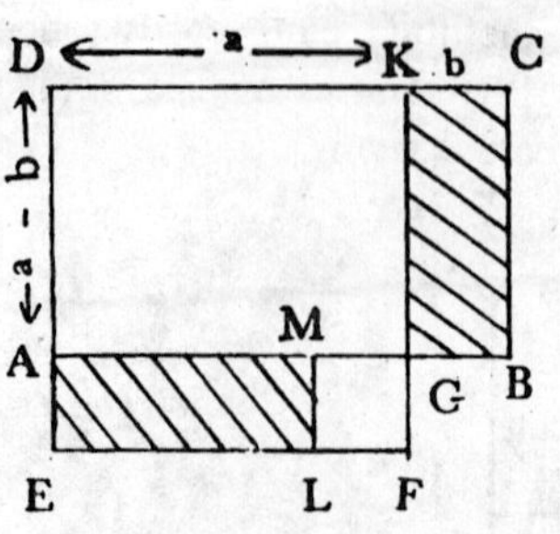

Fig. 16

9.8. The necessity for reducing fractions to a common denominator before they are added or subtracted is also explained

[1]Ibid p. 20.
[2]Ibid pp. 24-25.

concretely i.e. geometrically.[1] For example, taking $\frac{1}{4}$ and $\frac{1}{5}$, $\frac{1}{4}$ is one part obtained by splitting a strip of length 1 unit parallel to the breadth into 4 parts, whereas $\frac{1}{5}$ is to be obtained by splitting a similar strip into 5 equal parts. Now if the $\frac{1}{4}$ th part is again divided into 5 equal parts and the 1/5th part into 4 equal parts, these smaller strips of the two will be equal in size so that any number of strips from one can be counted together with, that is added to, any number of strips from the other.

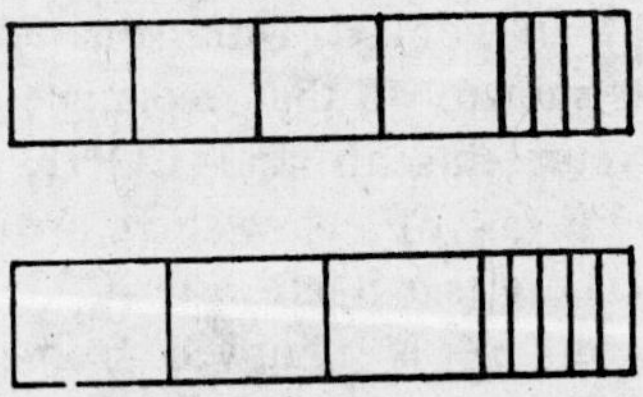

Fig. 17

9.9. The editors of the *Yuktibhāṣā* show how the multiplication of fractions can be demonstrated geometrically. The method, they say, is explained in another work (*granthāntara*) and is quite simple and easily understood from the attached figure.[2]

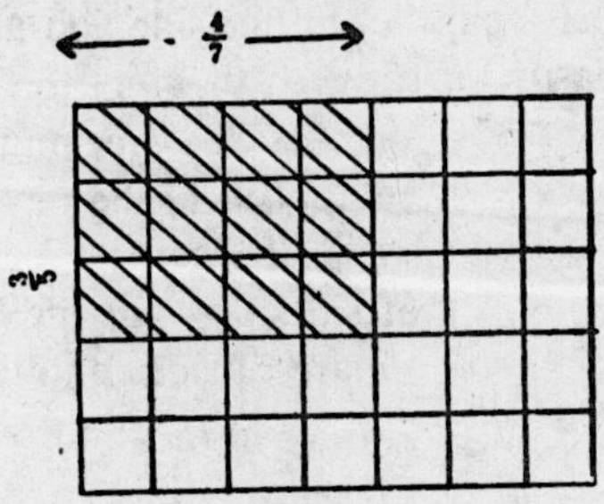

Fig. 18

The *Kriyākramakarī*, an elaborate commentary on the *Līlāvatī*, is another important work belonging to the Āryabhaṭa School. This two has a definite bias towards geometrical reasoning.

[1]Ibid p. 37.

[2]Ibid p. 41. footnote. One does not know what this other work is.

9.10. The demonstration for $(a + b)^2 = a^2 + b^2 + 2ab$ is the usual one. But for showing $(a + b)^2 = 4ab + (a-b)^2$, a slightly different procedure is adopted.[1]

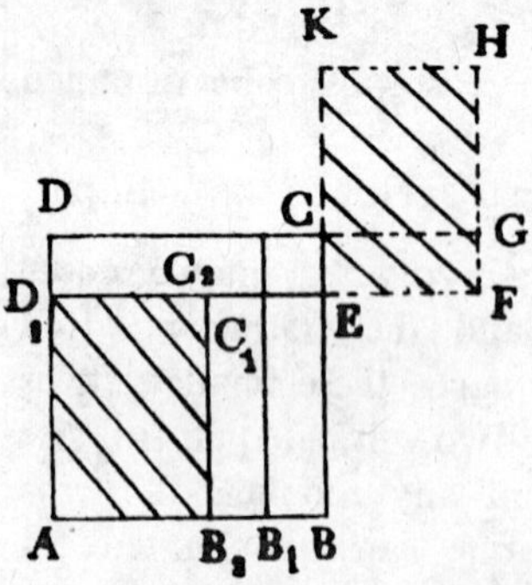

Fig. 19.

Let ABCD represent $(a + b)^2$. Let the figure be cut up as shown so that $AB_2 = a-b = D_1C_2$ and $B_2B_1 = C_2C_1 = b$. Remove the rectangle $AB_2C_2D_1$ and place it in contact with square CC_1 as shown, so that rectangle C_1G represents ab and CGHK represents $(a-b)^2$. Now there are four rectangles a b left after C G H $K = (a-b)^2$ is removed from A B C $D = (a+b)^2$. Arrange these to form the outline of a square. The sides of the square are $(a + b)$ and the sides of the inner square are $(a-b)$.

$$\therefore (a + b)^2 = 4ab + (a-b)^2.$$

Here is double proof, one by cutting up $(a+b)^2$ and the other by building up $(a + b)^2$ from 4 a b and $(a-b)^2$.

9.11. If a square A B C D of sides a is cut up into rectangles by a line EF parallel to one side and at a distance of b from it, it can easily be shown[2] that $a^2 = (a + b)(a-b) + b^2$. The rectangle AEFD can be applied to the rectangle F E B C, so as to get the rectangle F C L K of sides $(a + b)$ and $(a - b)$ with a square of sides b jutting out of it.

$$\therefore (a + b)(a - b) + b^2 = a^2.$$

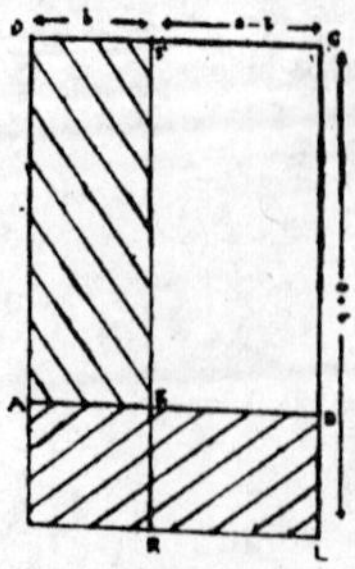

Fig. 20

9.12. Though this is proof enough for $(a + b)(a - b) = a^2 - b^2$ also, a separate proof[3] is given starting with a rectangle of

[1] *K.K.* p. 59.
[2] Ibid p. 60.
[3] Ibid p. 61.

sides $(a + b)$ and $(a - b)$. This can be cut up into 2 rectangles $a(a - b)$ and $b(a - b)$ by the line EF. Again $a(a - b)$ i.e. AEFD is to be cut up into a square of sides $(a - b)$ and a rectangle $b(a - b)$ by the line GH. Apply the rectangle AGHD to the square $(a - b)^2$, so that its long side coincides with GE. Now we get a square of sides a with a square of sides b unfilled one of its corners.

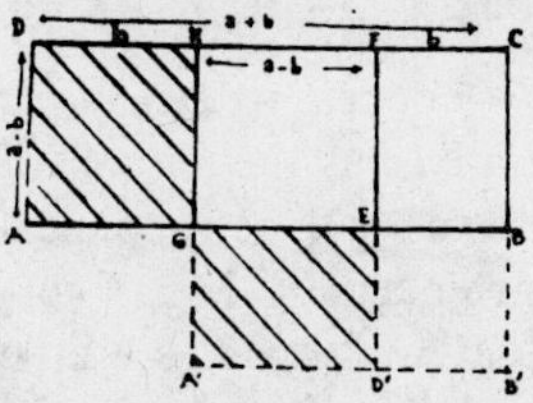

Fig. 21

$$\therefore (a + b)(a - b) = a^2 - b^2.$$

9.13. That $(a - b)^3 = a^3 + b^3 + 3a^2b + 3ab^2$ is demonstrated[1] by cutting up a cube of sides $a + b$ into 8 bits by planes parallel to the faces and alternatively by building up a cube of side $a + b$ from two cubes of sides a and b respectively and 3 rectangular parallelopipeds of sides a, b and $a + b$.

9.14. Another equality of the third degree demonstrated geometrically[2] is $a(a + i)(a - i) = a^3 - i^2 . a$

From a block of length $a + i$, breadth $= a - i$ and height $= a$, a piece of breadth i can be cut off from the length and joined to the breadth, so that both equal a. But at one corner a portion with a square base of sides i and height $= a$ will be left unfilled up. Hence the formula.

9.15. After explaining Bhāskara's rule for *Saṁkramaṇa* the *Kriyākramakarī* has an interesting passage.

अत्र राश्योर्योगभेदघातवर्गघनतन्मूलेषु द्वाभ्यां द्वाभ्यां विदिताभ्यां राशिद्वयानयनं एकविंशतिधा कार्यमित्युपदिष्टं चित्रभानुनाम्ना गणितगोलयुक्तिविदग्रेसरेण भूसूरोत्तमेन। तत्र दिङ्मात्रमस्माभिस्तदुपदेशवशादिह लिख्यते।

(p. 202)

[1]Ibid. pp. 92-94.
[2]Ibid. pp. 94-95.

(With two of the combinations of two quantities viz sum, difference, product, sum or difference of squares, sum or difference of cubes and the quantities themselves known the two quantities can be obtained. This can be done in 21 ways. Thus the Brāhman, Citrabhānu, by name, well-versed in the rationale of computations and astronomy has instructed. Instructed as I am by him, I write a little of it here.)

This Citrabhānu must be the same as is noticed in the *Keraḷa Sāhitya Caritram* of Ullur S. Parameśvara Iyer as the author of a commentary on Bhāravi's *Kirātārjunīya* and an astronomical work by name *Karaṇāmṛta.* His date is given as 706 M.E. corresponding to 1524-25 A.D. It is sad that we know so little about the mathematical achievements of this mathematician, held in such reverence by the author of the *Kriyākramakarī.*

Most of Citrabhānu's methods of calculation are algebraical. But characteristically enough, the logic of the procedure is often demonstrated graphically. The results so treated are:

1. When $a + b$ and $a^3 - b^3$ are known

$$\frac{4(a^3 - b^3) - (a - b)^3}{3(a + b)^2} = a - b, \text{ whence}$$

with $a + b$ known, a and b can be obtained by *saṁkramaṇa*[1].

$a^3 - b^3$ may be represented as a cube of side 'a' with a cube of side 'b' scooped out of one of its corners or, as the *Kriyākramakarī* puts it, as a floor $a - b$ thick with walls of the same thickness standing on two adjacent edges. The two walls, sides a and b and b and b, are to be separated and laid flat on the ground so that they form a rectangular block of sides $a + b$ and b. The floor, a square block of sides 'a', is then to be joined to this so as to be continuous with the larger wall. Then we get a square block of sides $a + b$ with a corner of sides a and b unfilled. Three of the four $(a^3 - b^3)$ blocks are to be treated similarly.

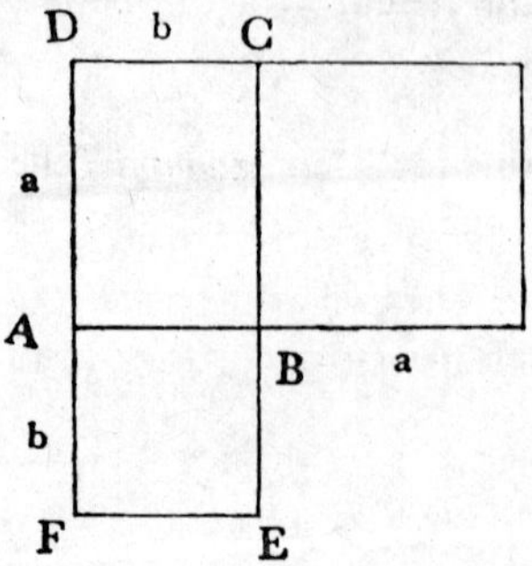

*Fig.*22

Then the empty corners will be

[1]Ibid p. 209.

three a. b blocks. The fourth $a^3 - b^3$ block is to be cut up so as to yield these three a. b blocks. The wall 'a' in length chopped just above the floor, yields one such block. The other wall, if chopped along with part of the floor, is another block of the same size, while the remainder of the floor is a rectangular block of the same size but with a square block of sides $(a - b)$ projecting at one corner. Since its thickness also $= a - b$ the projecting block is a cube of sides $(a - b)$. Thus, if a cube of sides $a - b$ is removed from $4(a^3 - b^3)$, the latter can be arranged as 3 square blocks of thickness $(a - b)$ and sides $(a + b)$.

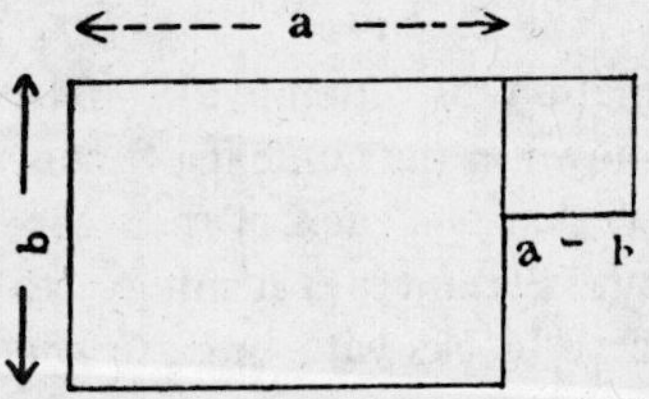

Fig. 23

Hence $$\frac{4(a^3 - b^3) - (a-b)^3}{3(a+b)^2} = a - b$$

Though the identity thus geometrically demonstrated is sound, since the equation is cubic and the quantity to be found out i.e. $(a - b)$ appears on both the sides, it does not help in the determination of $a - b$ except in some isolated cases. The remark applies also to the identities which follow.

2. When $a - b$ and $a^3 + b^3$ are known,

$$a + b = \left\{ 4(a^3 + b^3) - 3(a+b)(a-b)^2 \right\}^{\frac{1}{3}}$$

[1]Here $a^3 + b^3$ is representable as a cube of side 'a' with a cube of side 'b' placed on top of it (if a is larger then b), with two of its faces lying along the same planes as two faces of the lower cube. The jutting portion of the lower cube, which will be a gnomon of thickness $a - b$, height a and combined length of limbs $(a + b)$, is sliced off. A second $(a^3 + b^3)$ set is similarly treated. Then the two gnomons are placed on the 3rd and the fourth $a^3 + b^3$ blocks so that they fit around the top b^3 blocks, but leave a gnomon of height $(a - b)$ projecting. This projecting portion is cut off. Thus we get two pillars on square bases of sides 'a' and height $(a + b)$ and two pillars of sides 'b' of the same height. These four can be put together to form a cube of side $(a + b)$ with a square prism of length $(a + b)$ and sides

[1]Ibid p, 213.

(a — b) projecting out of it. The unused part of the gnomons of the first two ($a^3 + b^3$) blocks also, when the limbs are separated and joined lengthwise, yield similar prisms. Hence four a^3+b^3 block yield a cube of side (a + b) along with three prisms of length (a + b) on a square base of sides (a — b).

$$\text{Hence } (a + b)^3 = 4\,(a^3 + b\,) - 3\,(a + b)\,(a - b)^2.$$

3. [1]When a—b and a^3-b^3 are known $\dfrac{a^3-b^3-(a-b)^3}{3\,(a-b)} = ab$

For this also a geometrical explanation is attempted. "The rationale of this has been already shown in the context of cubing. When two cubes are placed so that one each of their corner-edges touch each other (without the faces touching), to their three sides are to be attached three blocks with breadth and height equal to the two quantities and length equal to their sum. As the cube of side equal to their sum results, when these 5 blocks are joined properly, the difference of the cubes (is obtained) when the smaller cube is removed. The two cubes are separated. Then two bits as also three products of the bits remain. Hence their sum is obtained by dividing by thrice their difference." The latter part of the passage looksconfused. The demonstration can actually be the same as the one given for the next equality.

4. When ab and a^3-b^3 are known $\dfrac{a^3-b^3-(a-b)^3}{3\,a\,b} = a-b$

[2]The rationale is implied in the graphical representation of $(a + b)^3$. In the cubical block there are 5 blocks viz., the cubes of the two parts and rectangular blocks of sides equal to the two parts and the sum of the parts respectively. If now we make the whole cube represent a^3 and remove one of the cubical blocks contained in it, the remaining blocks will represent $a^3 - b^3$ and the cubical one among these will represent $(a-b)^3$.

$$\text{Hence } a^3-b^3 = (a-b)^3 + 3ab\,(a-b) \text{ or}$$

$$a-b = \frac{a^2-b^3-(a-b)^3}{3\,ab}$$

[1]Ibid pp. 216-17.
[2]Ibid pp. 221-22.

or to get $a-b$, a^3-b^3 is to be divided by $3ab$ after being diminished by the cube of the quotient so obtained.

5. [1]When a^2+b^2 and a^3+b^3 are known, the quotient got by dividing a^3+b^3 by a^2+b^2 is the smaller of the two quantities a, b. The remainder when divided by a^2+b^2 diminished by the square of the above quotient, gives $a-b$.

Let a piece of the height of the smaller quantity (b) be cut off from the larger cube (a^3) and let this be joined to the smaller cube b^3. Then we get a block answering to $(a^2+b^2).b$. The remainder block is $a^2(a-b)$. Hence the integer got by dividing a^3+b^3 by a^2+b^2 is b. The remainder when divided by a^2 i. e. a^2+b^2 diminished by the square of the above quotient, yields $(a-b)$. Except when $a-b=1$, the quotient of division of a^3+b^3 by a^2+b^2 should not be taken as the largest but the one after subtracting which from a^2+b^2 it is possible to divide the remainder by the difference without leaving a remainder.

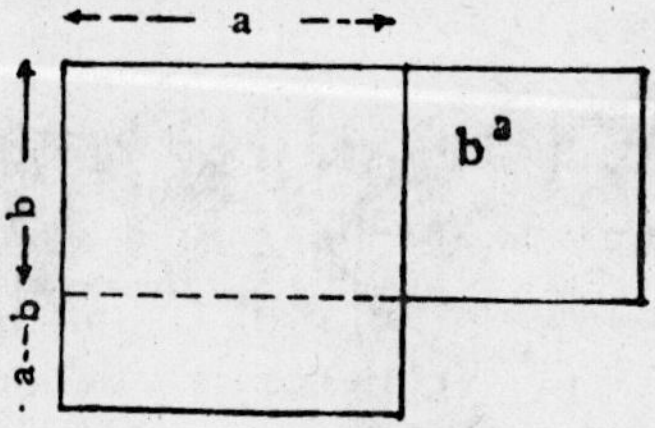

Fig. 24

6. [2]When a^2+b^2 and a^3-b^3 are known

$$\frac{2(a^3-b^3)+(a-b)^3}{3(a^2+b^2)}=a-b.$$

From the difference of the cubes representable as a platform (or floor, the Saṃskṛt word is *kuṭṭima*) with walls at two adjacent edges i. e. as a cube of sides 'a' with a smaller cube of side 'b' scooped out at one corner, the *kuṭṭima* which is $(a-b)$ in thickness is sliced off. The two walls also should be separated. These

[1]Ibid. pp. 222-24.
[2]Ibid pp. 224-25.

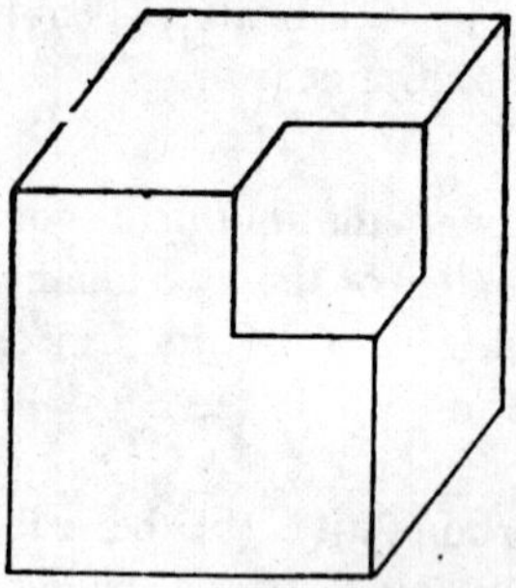

Fig. 25

will be a rectangular block of sides a and b and a square block of sides b, both having the same thickness (a—b).

If the *kuṭṭima* and the square wall are joined together we get a larger block of area of section $a^2 + b^2$ and thickness a—b. A second (a^3-b^3) block is subjected to the same treatment. Then we get four blocks corresponding altogether to $2(a^2+b^2)$ (a—b) and 2 a b (a—b).

Now $2ab = a^2 + b^2 - (a-b)^2$

$\therefore 2ab(a-b) = (a^2 + b^2)(a-b) - (a-b)^3$

Hence $2(a^3 + b^3) = 3(a^2 + b^2)(a-b) - (a-b)^3$

or $2(a^3 + b^3)$ is exactly divisible by $(a^2 + b^2)$ 3, if $(a-b)^3$ is added to it. And the quotient is (a—b). Hence the rule for finding (a—b), viz., divide $2(a^3 + b^3)$ by $3(a^2 + b^2)$, add the cube of the largest integer quotient so obtained to $2(a^3 + b^3)$ to make it exactly divisible. The quotient is (a—b).

Here there is an effective mixture of geometrical and algebraical reasoning.

7. [1]When a^2-b^2 and a^3+b^3 are known, the quotient obtained by dividing $a^3 + b^3$ by $a^2 - b^2$ is the larger number, a. The remainder divided by the above quotient diminished by a^2-b^2, yields a + b.

If the smaller of the cubes a^3 and b^3 is placed on top of the larger, the projecting part of the larger block will be a^2-b^2 in sectional area and 'a' in height, while the other part will be a prism on a square base of side b and height a + b

Hence $a^3 + b^3 = b^2(a + b) + a(a^2-b^2)$

$$\text{or } a = \frac{a^3 + b^3 - b^2(a + b)}{a^2-b^2}$$

Hence the rule.

[1]Ibid pp. 226-27.

9.16. The lemma in the Lilāvatī[1] that a^2+b^2-1 will be a perfect square, if $a = 8p^4 + 1$ and $b = 8p^3$, (where p is any arbitrary number), is furnished with an interesting graphical proof by the *Kriyā-kramakarī*[2]

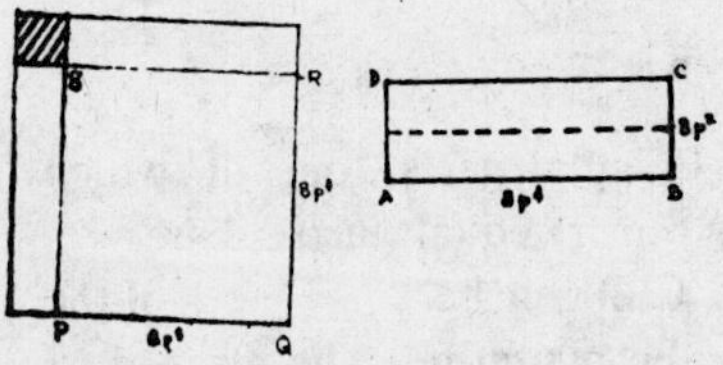

Fig. 26

$(8p^4)$ can be represented as a square PQRS of sides $8p^4$, and $(8p^3)^2 = 8p^4 . 8p^2$ as a rectangle ABCD of sides $8p^4$ and $8p^2$. ABCD is divided into two equal parts by a line parallel to the long sides, when the sides of these strips will be $8p^4$ and $4p^2$. Apply these strips to the square PQRS so as to form a square of sides $8p^4 + 4p^2$ but with a square of sides $4p^2$ unfilled up at one corner.

But $a^2 = (8p^4 + 1)^2 = (8p^4)^2 + 16p^4 + 1$ Hence $16p^4 + 1$ has yet to be used.

The part $16p^4 = (4p^2)^2$ can be used to fill up the empty corner in the larger square made up of $(8p^4)^2$ and $(8p^3)^2$; i. e. $(8p^4 + 1)^2 + (8p^3)^2$, can be converted into a square, with 1 only not included. Hence when 1 is subtracted from $a^2 + b^2$ i.e. from $(8p^4 + 1)^2 + (8p^3)^2$. the difference can be represented as a complete square.

The same demonstration with slight changes holds for a^2-b^2-1.

9.17. An equally interesting graphic demonstration is the proof given by the *Kriyākramakarī*[3] for the expression $S_n = \dfrac{ar_n - a}{r-1}$ for the sum of a geometrical progression. The proof is given for a particular series, the geometrical progression with 4 as the common ratio. But with 4 changed to r it is applicable to any geometrical progression.

Let a long rectangular strip ABCD represent the (n + 1)th term of the progression.

[1]V. 63.
[2]*K.K.* 254
[3]Ibid p. 458.

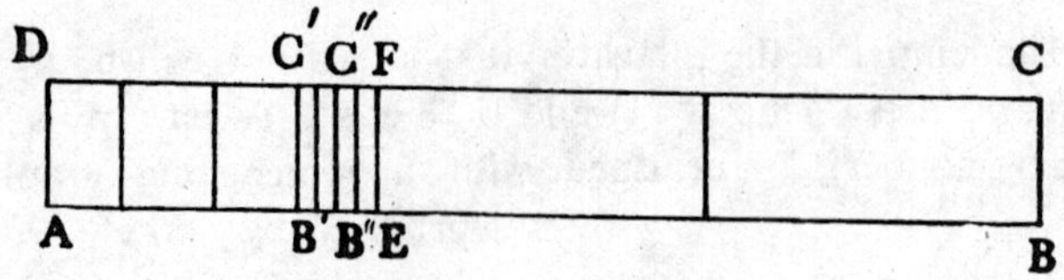

Fig. 27

Let it be divided into 3 (i.e. $r-1$) equal parts, one of which AEFD is again divided into 4 (i.e. r) equal parts. Three of these smaller divisions i.e. AB′C′D will be $\frac{3}{4} \cdot \frac{1}{3} = \frac{1}{4}$ of the $(n+1)$th term and hence equal to the nth term. The procedure is repeated with the remainder B′ E F C′ which is equal to one third of the nth term. Then we get the $(n-1)^{th}$ term. The remainder when treated similarly will yield the $(n-2)^{th}$ term. This process is repeated till the first term is reached. The part left over will then be $\frac{1}{3}$ of the first term.

Hence $\frac{1}{3}$ the $(n+1)^{th}$ **term** = the sum of n terms + $\frac{1}{3}$ lst term

Whence, since $\frac{1}{3}$ is $\dfrac{1}{r-1}$

$$S_n + \frac{a}{r-1} = \frac{1}{r-1} . a . r^n$$

$$\text{Or, } S_n = \frac{a . r^n - a}{r-1}$$

9.18. The diagrammatical representation of a product as an area is used in an unexpected context, i.e. while explaining the famous problem of the *uḍḍīnamāna* (the height of flight). Two monkeys on the top of a palm want to reach a lake some distance from the palm. One of them rises vertically up into the air for some time and then swoops diagonally into the lake. The other climbs down the trunk of the tree and from its root proceeds to the lake. If the distance traversed by the two is the same, what is the height of flight?

Deriving the expression for the unknown height of flight, x, the commentator arrives at the equality.

$$\frac{b\,(h + x)}{b + k + p} = x$$

Or $x(b + k + p) = b(h + x)$

from which $x = \dfrac{bh}{k + p}$

For these final steps the *Kriyākramkari* appends a diagrammatic corroboration. $x(b + k + p)$ can be represented as a rectangle ABCD of width x and length $(b + k + p)$.

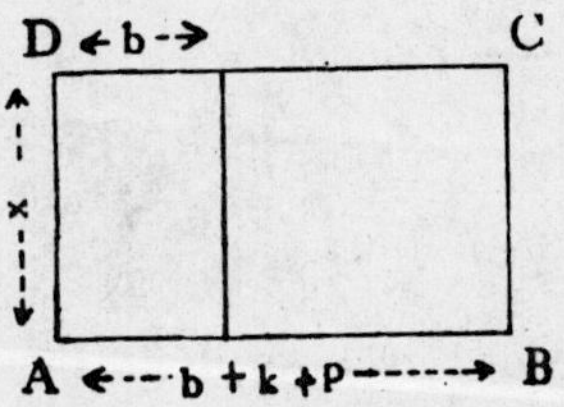

Fig. 28

To remove bx from this, a part of the rectangle, 'b' in length, is cut off from ABCD. Then the length of the remainder is $k + p$ and its width is x.

$$\therefore x = \frac{\text{area of the remainder rectangle}}{k + p}$$

This is an example of an exercise in diagrammatic imagination which does not seem to further the understanding or establishment of the expression to any appreciable extent. The school to which the *Kriyākramakarī* belongs seems to have formed the habit of finding satisfying corroboration in practical demonstration.

9.19. A sign of a similar dependence on diagrams for understanding and corroborating algebraical identities is found in *'Dhavaḷā*, Vīrasena's commentary on the *Ṣaṭkhaṇḍāgama*. Says Vīrasena

द्विभागाभ्यधिकसर्वजीवराशिना सर्वजीवराश्युपरिमवर्गे भागे हृते किमागच्छति ? त्रिभागहीनसर्वजीवराशिरागच्छति । केन कारणेन ? सर्वजीवराशिवर्गक्षेत्रं पूर्वापरायामेन त्रीणि खण्डानि कृत्वा तत्रैकखण्डं गृहीत्वा खंडं कृत्वा संधिते सर्वजीवराशिद्विभागविस्तरि भवति । भागायामक्षेत्रं भवति ।

(*Ṣaṭkhaṇḍāgama*, Vol. III. p. 44)

(When the square of the total number of creatures is divided by that total number increased by its own half what is the result? The result is the total number of creatures diminished by one-third of itself. Why? The figure representing the square of the number of creatures is to be divided into 3 parts by lines running from east to west. One of these parts is divided into 2 equal parts and joined to the other two parts; the width (of the resulting figure) will be $\frac{2}{3}$ the total number of creatures. This is the *bhāgāyāmakṣetra*.)

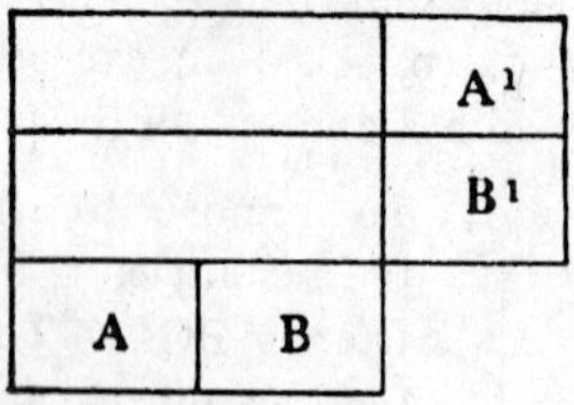

Fig. 29

If the square of the *jīvarāśi* is to be divided by $1\frac{1}{3}$ of itself, the square should be divided into 4 equal parts. One part is then divided into 3 equal parts and the parts joined to the remaining three parts. And so on. i.e., if A $(=a^2)$ is to be divided by $(1+\frac{1}{n})$ of itself, the square representing A should be divided into $(n+1)$ equal parts, one $(n+1)$th part again divided into n equal parts, and the parts joined to the remaining $(n+1)$th parts to form a rectangle of length $a.(1+\frac{1}{n})$ and breadth $a.(1-\frac{1}{1+n})$.

Then the breadth gives the required quotient.

With hardly a mathematical work of the Jainas left to us, it is very difficult to know how far they had yoked geometry to algebra. Still one can say that it is not unlikely that the Ārya-bhaṭa School derived its love of geometrical algebra from the Jainas.

9.20.1. *Śreḍhīkṣetras*

We have already seen how the *Kriyākramakarī* proves the formula for the sum of a geometrical progression diagrammatically. The method of diagrammatic representation was applied more extensively to problems connected with Arithmetical progressions (A. P.). These representations go by the name of *śreḍhīkṣetras* (figures of series). Mention and examples of *śreḍhīkṣetras* are of late occurrence. But it is quite likely that the conception and use of such diagrams is quite old. For, in the *Tantrabhāṣya*, Bhāskara I's commentary on the Āryabhāṭīya, mathematics is divided into two sections, *Rāśigaṇita* (mathematics of numbers) and *Kṣetragaṇita* (mathematics of figures), and then Bhāskara says

अनुपातकुट्टाकारादयो गणितविशेषाः राशिगणे (णिते ?) ऽभिहिताः श्रेढीच्छायादयो क्षेत्रगणिते

(*Bhāskarakṛtam Āryabhaṭiyabhāṣyam p.* 55)

(Mathematical topics like proportion and *kuṭṭākāra* (indeterminate analysis) are enumerated in the mathematics of numbers; series, shadow problems etc. in the mathematics of figures.)

Here *śreḍhī* or series is included under *Kṣetragaṇita* or geometry. Does this mean that series originally formed part of geometry, the mathematics of figures? Āryabhaṭa I uses the terms *citighana*, *vargacitighana* and *ghanacitighana* for the sum of triangular numbers, the sum of the squares of the natural numbers and the sum of the cubes of the natural numbers respectively (*citi* means a pile and *ghana* means cubic contents). These terms may be explained only if we assume that Āryabhaṭa, studied these series in relation to piles. And this is not improbable, since mathematics first developed in India in connection with the construction of *vedis*, and slanting pile-like *vedis* (e.g. the *Samūhya* and *Paricāyya* are included among these.

9.20.2. Yet, except for this hint, we do not find attempts at the diagrammatical representation of *śreḍhīs* in the earlier mathematical works. The treatment of series in Mahāvīra's *Gaṇitasārasaṃgraha* and the *Bakhshali Manuscript* alone confronts us with an unusual feature, namely series with fractional numbers of terms. In the *Pāṭīgaṇita* of Śrīdhara and the *Gaṇitakaumudī* of Nārāyaṇa, on the other hand, besides series with fractional periods we come across ones with negative periods, negative sums and sums equal to zero. All these occur in connection with *śreḍhīkṣetras* only. Nārāyaṇa's description of *śreḍhīkṣetras* runs

आदिश्चयदलहीनो वदनं पदचयवधः सवदनो भूः।
गच्छो लम्बो गणितं श्रेढीगणितेन तुल्यं स्यात् ॥

अवलम्बखण्डगुणितश्चयः स्ववदनेन संयुतस्तद्भूः ।
ऋणगे वदने तु मिथो भुजौ समाक्रम्य वर्धेते ॥
अधरोत्तरे भवेतां त्र्यस्रे भूवदन-भूमिके स्वर्णे ।
विवदनकुहृते कुमुखे लम्बघ्नौ त्र्यस्रयोर्लम्बौ ॥
तद्गुणितयोश्च विवरं श्रेढीगणितेन वा तुल्यम् ।

(*G. K., Ks, Vya.*, 73-75$\frac{1}{2}$)

(The first term (of the series) diminished by half the common difference (C. D.) is the face, the product of the period and C. D. increased by the face is the base; the period is the altitude and the area is the sum of the series. The fraction of the altitude multiplied by the C. D. and combined with its own face is the base (of any segment of the trapezium.) If the face is negative the two flanks will cross each other and grow. Then there will be two triangles one positive and one negative with the base and the face as the bases. The base and the face multiplied by the altitude and divided by the base minus the face are the respective altitudes of the triangles. The difference of their areas will be equal to the sum of the series). The first verse represents an A. P. as a trapezium with altitude equal to the period of the A. P. But instead of making the face (or the smaller parallel side) equal to the first term of the series, $\left(a - \frac{d}{2}\right)$ is made the face and $\left(n\,d + a - \frac{d}{2}\right)$ the base (a is the first term, d the C. D., and n the number of terms of the A. P.). Thus it is possible to have the face negative even when the first term of the A.P. is not negative. The second verse gives the method for calculating the base at any intermediate position on the altitude, i. e. when the altitude is any fraction of the whole altitude. Since this fraction of the altitude or period need not always work out to be a whole number, summation of series with fractional number of terms becomes natural. Next we are given a picture of the *śreḍhīkṣetra* when

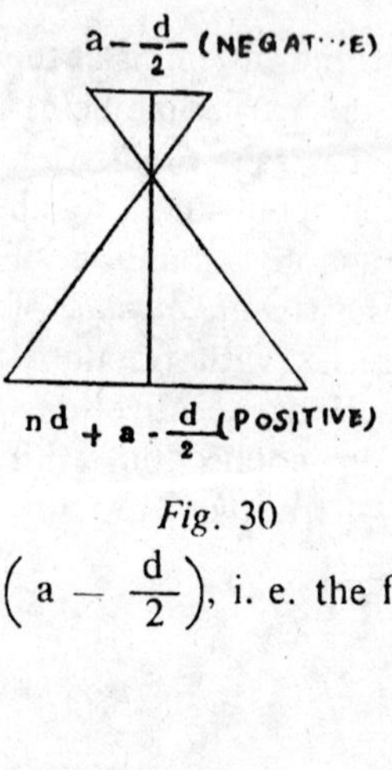

Fig. 30

$\left(a - \frac{d}{2}\right)$, i. e. the face turns out negative and we are told

how to calculate the altitudes of the two triangles making up the whole *śreḍhīkṣetra*. The formula given is:

$$h_1 = \frac{\text{face}}{\text{base} - \text{face}} \text{ . whole altitude}$$

$$\text{and} \quad h_2 = \frac{\text{base}}{\text{base} - \text{face}} \text{. whole altitude}$$

(When the two triangles are considered as similar triangles, the expressions for h_1 and h_2 will be

$$h_1 = \frac{\text{face}}{\text{base} + \text{face}} (h_1 + h_2)$$

$$\text{and} \quad h_2 \frac{\text{base}}{\text{base} + \text{face}} (h_1 + h_2)$$

But we have to remember that the face is a negative quantity here and that is why Nārāyaṇa makes the base minus the face the denominator.)

The difference of the areas of the two triangles will be the sum of the A. P.

Śrīdhara's account (*Pāṭīgaṇita* V. 79-85) is similar to this, except that he makes $a - \frac{d}{2}$ the base of the figure, so that it is narrower at the base and wider at the top like a cup (*śarāva*). Also for the actual construction he recommends the construction of the series figure (*śreḍhīkṣetra*) for unit altitude. (This is called *hāstikakṣetra* since the unit used is the cubit, *hasta* or *kara*). Then the face of the actual series figure with altitude = the number of terms in the series is to be calculated using the principle of proportionate increase.

As illustration, Nārāyaṇa, as also Śrīdhara's commentator gives a number of A. P. s with drawings of their *śreḍhīkṣetras* and calculations of the elements of these figures. One of these has the first term $a = \frac{1}{2}$ the C. D., $d = 3$ and the period, $n = 3\frac{1}{2}$ (*G. K. Ks. Vya.* p. 88, V. 62). The face of the *śreḍhīkṣetra* is $\frac{1}{2} - \frac{3}{2} = -1$.

The figure then takes the form of two inverted triangles joined at

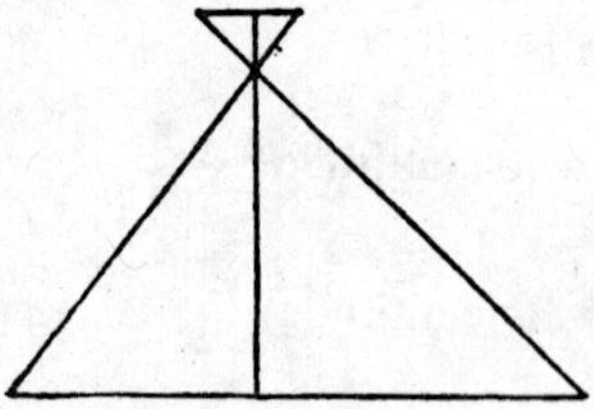

Fig. 31

their apexes and the altitudes of the two triangles are $\frac{1}{3}$ and 3 respectively. The next example (p. 89. V. 63) has $a = 3$, $d = 7$ and $n = \frac{1}{7}$. The face of the *śreḍhīkṣetra* is here $3 - \frac{7}{2} = \frac{-1}{2}$. The base $=$ $= nd + \text{face} = 1 - \frac{1}{2} = \frac{1}{2}$ and the altitudes of the two triangles are $\frac{1}{14}$ and $\frac{1}{14}$. Hence the sum of the A. P. is zero. (The figure is not shown).

The next two examples (p. 90., V. 64 & 65) have negative periods. The conception of a series with negative period is even stranger and more difficult than of one with fractional period.

9.20.3. The inverse process of converting a trapezium into an A. P. is detailed in

लम्बोद्धृतविमुखभूः प्रचयश्चयदलयुतं वदनमादिः ।
लम्बो गच्छः श्रेढीगणितं गणितेन तुल्यं स्यात् ॥
क्षयगे वदने तु समो मध्यमलम्बोऽवलम्बकाभ्यां चेत् ।
आदिचयोत्पत्तिः स्यान्न चान्यथा विषमचतुरस्रे ॥

(*G.K.*, *Ks. Vya.*, pp. 76-77)

(The base dimininished by the face and divided by the altitude is the C. D., the face combined with half the C. D. is the first term: the altitude is the period and the area is the sum of the A. P. If the face is negative, the first term and the C. D. can be obtained only if the altitude at the middle is equal to the altitudes at the vertices, not otherwise in a quadrilateral of unequal sides.) The restriction in the second verse seems unnecessary, since, in any case, the quadrilateral has to be a trapezium, if it is to be represented as an A. P. In his example for a *viṣama* quadrilateral (*G. K.*, *Ks. Vya.* p. 93) the author himself calculates the area as if the quadrilateral is a trapezium.

In connection with this reverse process too, Nārāyaṇa has, it is noteworthy, an example where the area of the trapezium and hence the sum of the A. P. is zero and the figure takes the form of two equal triangles joined with their apexes coinciding, the whole somewhat resembling the wave pattern traced out by an alternating current.

9.20.4. The nature and the use of *śreḍhīkṣetras* after Nārāyaṇa Paṇḍita seem to be altogether different. This use is met with chiefly in Nīlakaṇṭha's *Āryabhaṭīyabhāṣya* and the *Kriyākrama-karī*, written probably by a pupil of Nīlakaṇṭha. The former introduces *śreḍhīkṣetras* to establish the correctness of the summation formulae given by Āryabhaṭa. Commenting on *Gaṇita-pāda*, verse 19,

इष्टं व्येकं दलितं समूलमुत्तरगुणं समुखमध्यम् ।
इष्टगुणितमिष्टधनं त्वथवाद्यन्तं पदार्धहतम् ॥

Nīlakaṇṭha explains how a *śreḍhīkṣetra* is to be constructed. A rectangle is drawn with one side containing as many units as the period n and the other side as many units as the last term, *l*. Divide the side = n into n equal parts by lines parallel to the other side and the side = l into l equal parts similarly. Now the figure is divided into strips containing l small squares each. In one of the outermost of these strips keep only as many small squares as there are units in the first term, a, and wipe off the remainder. In the second strip keep a + d squares and wipe off the rest. In the third keep a + 2 d and so on, till the last row is reached in which no square is to be erased. Now the *śreḍhīkṣetra* is complete and looks like A B C D in the figure.

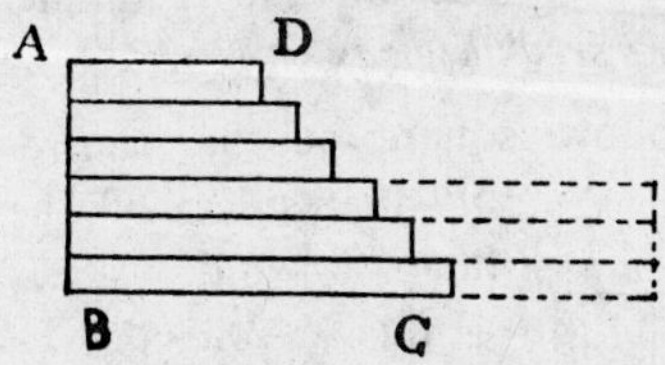

Fig. 32

To get the sum, lift the first strip and join it with the last strip so that their lengths lie along the same line. The length of the strip is now l + a. Join the second strip with the (n—1)th strip. Its length also will be l + a. In this way join one strip from the upper half to one strip in the lower half and complete the process so as to leave $\frac{n}{2}$ strips of length a + l. Then the area of the rectangle thus formed is $(a + l) \cdot \frac{n}{2}$. This particular method is applicable only when n is even. When n is odd two *śreḍhīkṣetras* can be joined inverted so as to yield a rectangle of sides (l+a) and n.

9.20.5. Āryabhaṭa's expression for the period n of an A.P. is :

$$n = \frac{1}{2}\left\{\frac{\sqrt{8\,d\,s + (d \sim 2\,a)^2} - 2a}{d} + 1\right\}$$

where d is the C D, s the sum and a the first term of the A. P. (*Gaṇitapāda*, 20). Nīlakaṇṭha gives a simple and convincing geometrical proof for this with the help of *śreḍhīkṣetras*. Since the expression contains the term 8 d s, take 8 d *śreḍhīkṣetras*. Combine pairs of such figures inverted so as to form rectangles of sides a + l units (known) and n units (to be found out). There will be $\frac{8}{2}$ d, i.e., 4 d such rectangles. Join d of such rectangles together by their known sides so that the unknown sides lie along the same line as shown in the figure (Fig. 33).

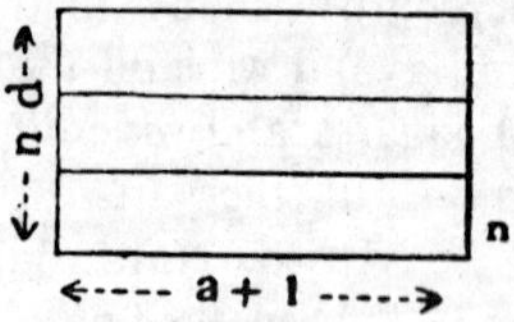

Fig. 33

Thus we get 4 big rectangles of sides n d and (a + l). Now join these 4 equal rectangles as shown in Fig. 34 below to form a square with an empty square at its centre. The side of the outer square will be the sum of the sides of the rectangle formed by combining d pairs of *śreḍhīkṣetras*, i. e. n d + a + l.

The side of the hollow square at the centre is the difference of the sides of the rectangle, i.e., (a + l)—n d., i.e., 2 a + (n—1) d — nd, i.e. 2 a∼d. Hence 8 d *śreḍhīkṣetras* together with a square of side (2 a ∼ d) form a square of side (nd + a + l), i.e.

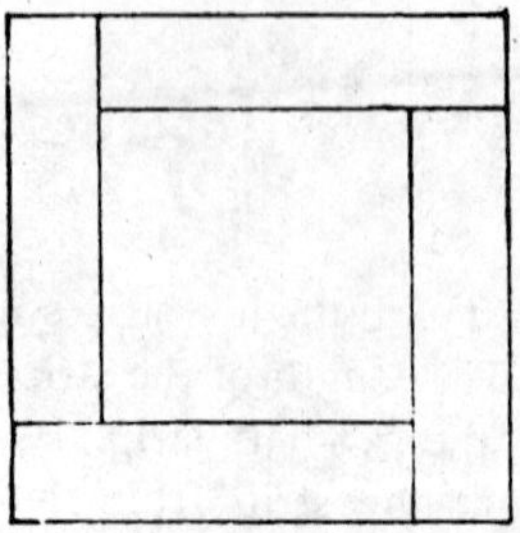
Fig. 34

$n\,d + 2\,a + (n-1)\,d = 2\,nd + 2a - d$

i.e. $2 \times nd + 2a - d$

$= \sqrt{8\,d\,s + (2\,a \sim d)^2}$

i.e. $n = \frac{1}{2}\left\{\frac{\sqrt{8\,d\,s + (2\,a \sim d)^2} - 2\,a}{d} + 1\right\}$

The author of the *Kriyākramakarī* gives the same proof after remarking that the formula as given by Śrīdhara and Parameśvara agrees with this, while Bhāskara II's expression is

$$\frac{\sqrt{2 . d . s + \left(\frac{d}{2} \sim a\right)^2} - a + \frac{d}{2}}{d}$$

The difference is because the former mathematicians arrived at the expression geometrically using 8 d *śreḍhīkṣetras,* while for demonstrating Bhāskara's expression, 2 × d *śreḍhīkṣetras* are used. The method is unsuitable when 8 d is an odd number, a fraction, or a negative number.

9.20.6. To prove that the sum of the first n triangular numbers, i.e. the *citighana,* is $\frac{n(n+1)(n+2)}{6}$ or $\frac{(n+1)^3 - (n+1)}{6}$ (*Gaṇitapāda,* 21) the same tool is used. The sum of the triangular numbers is Σs_n, i.e. $s_1 + s_2 + \ldots\ldots\ldots\ldots + s_n$, where s_n is the sum of n natural numbers, i.e., if we construct *śreḍhīkṣetras* for all the different sums involved, we will have n such figures all similar but gradually increasing in size. Taking 6 such sets Nīlakaṇṭha shows how a rectangular block of sides n, (n + 1) and (n + 2) can be constructed out of these.

Out of the 6 *śreḍhīkṣetras* representing s_n three rectangular slabs of sides n, n + 1 and unit thickness can be built up. One of these slabs is placed flat on the ground and the second is stood on its thickness with its breadth vertical and length coinciding with the length of the first slab. The third slab is now stood on its thickness with the breadth vertical and length coinciding with the line formed by the breadth of the first and the thickness of the second. The two will then coincide.

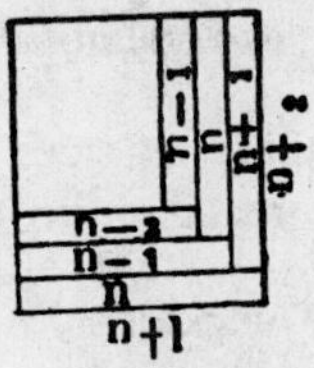

Fig. 35

Now we have a floor of sides n and (n + 1) and unit thickness with walls of unit thickness and height equal to n externally and (n − 1) internally standing at two adjacent sides. That is, the cuboid with its unfilled part will measure (n + 2), (n + 1) and n externally. Now three rectangles are formed from the 6 *śreḍhīkṣetras* representing s_{n-1}. These will have length n and breadth (n–1). Two of these are stood on their thicknesses with their breadth vertical and touching the internal surface of the walls of the cuboid already formed. The third one is placed flat on top of the floor of the cuboid. Now the walls and floor of the cuboid are 2 units in thickness and the hollow part is 'n' in length, (n–1) in breadth and (n–2) in height. Similarly if the remaining sets 6 s_{n-2}, 6 s_{n-3}, . . . 6 s_1 are subjected to the same treatment, the hollow in the cuboid will be completely filled up and no slab will be left over. That is, a rectangular block of sides n, (n + 1), (n + 2) can be built out of 6 Σs_n

$$\therefore \Sigma s_n = \frac{n(n+1)(n+2)}{6}$$

The conversion of this formula into the alternative form $\frac{(n+1)^3 - (n+1)}{6}$ is also effected without the help of algebra by cutting off a lamina 1 unit thick from one end of the above block perpendicular to the longest side (n + 2) and using it to increase the height of the block at one end by one unit. But, since the lamina is only (n + 1). n . 1, a portion 1 unit wide at the other end will fail to acquire the increased height n + 1 i.e. a rod 1 unit square in cross section and (n+1) units long is wanting to make the block a right cube of side (n + 1). Hence the volume of the block is $(n+1)^3-(n+1).\ 1^2$. Hence the expression.

The equality $\Sigma n^2 = \frac{n\,.\,(n+1)(2n+1)}{6}$ is also demonstrated similarly by treating 6 Σn^2 as 6 sets of n squares of sides 1, 2,..............n, converting these into 3 sets of n rectangles of length 2, 4,. . . . . .2 (n—1), 2n, and breadth 1, 2, 3, ...n—1, n and with these 3 sets building up a solid rectangular block of sides n, (n+1) and (2n+1).

9.20.7. For $\Sigma n^3 = \left\{\frac{n(n+1)}{2}\right\}^2$ a slightly modified method is used. Since $\frac{n(n+1)}{2}$ is the sum of the natural numbers,

$\left\{\frac{n(n+1)}{2}\right\}^2$ can be represented as a square block of side $\frac{n(n+1)}{2}$ each and thickness one unit. Cut off a gnomon of width n units from this block, which is then, in its turn, to be cut into small blocks beginning from the corner. The block at the corner is a square of side n. The remaining blocks on either side are to have one side decreased by 1 unit progressively, i.e. that side will be respectively (n—1), (n—2) ....1. Since the side of the original large block is the sum of the natural numbers, the gnomon will be finished by the time the block, 1 unit in width, is reached on both sides. Now keeping the first square block of side n apart, we have two sets of (n—1) rectangular blocks, each one unit in thickness and n units in length, but with the breadths gradually decreasing by one unit from (n—1) units to 1 unit. The first block from the first set of width (n—1) units is joined with the last block of width 1 unit from the second set to get a square of sides n units. Similarly the second from the first set, of width (n—2) units, is joined with the last but one in the second set of width 2 units to yield a square of sides n, and so on. Thus (n—1) square blocks of sides n units and thickness 1 unit are obtained. Now these (n—1) blocks are arranged on top of the square at the corner of the gnomon. The thickness or the height of the pile thus formed will be n units, so that a cube of sides n results.

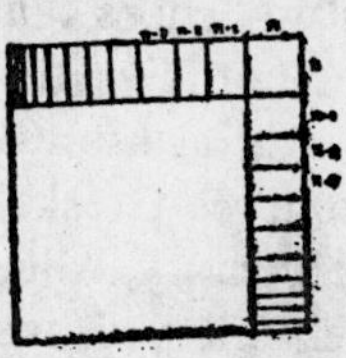

Fig. 36

By cutting off successive gnomons from the remainder of the original block (the side of the remainder block is s_{n-1}) of widths (n—1), (n—2) ...... 1, we can similarly build up cubes of sides (n—1), (n—2) ...... 1. That is, the set of n cubes of sides n, (n—1), (n—2) ...... 1 can be built out of the block of sides s_n, s_n and 1 unit,

$$\text{i.e. } \Sigma n^3 = \frac{n(n+1)}{2} \cdot \frac{n(n+1)}{2} \cdot 1$$

$$= \left\{\frac{n(n+1)}{2}\right\}^2$$

The history of this formula and its proof in the West is interesting in this connection.[1] Nicomachus, about A.D. 100, notes that the series of the odd natural numbers yields the cubes of the natural numbers, when its successive terms are grouped together in groups of 1, 2, 3, . . . terms, i.e. as 1, (3 + 5), (7 + 9 + 11) . . . . From this the expression for the sum of the natural cubes can be easily derived. But Nicomachus himself does not give the formula, though it was known to the Roman Agrimensores, who derived all his mathematical knowledge from the Greeks. Al-Karkhi, the Arabian algebraist of the 11th century, who, according to Sir Thomas Heath, follows Greek methods as opposed to Indian methods, proves this result with the help of a figure with gnomons in it. Such geometrical algebra is "distinctively Greek", adds Health. Whether the Arabian mathematician derived his proof from the Greeks, or invented it himself, or was influenced by Indian mathematics, we have to accept the fact that in India there was a school of mathematics with a distinct bias towards geometrical demonstration for establishing mathematical truths—the school of Āryabhaṭa.

9.20.8. The demonstrations of the above equalities as given in the *Kriyākramakarī* are substantially thesame. Only, after the demonstration for the equality

$$\Sigma s_n = \frac{n(n+1)(n+2)}{6}$$

i.e. for the *saṃkalita-saṃkalita,* the commentator says

गच्छाद्येकोत्तराङ्कानां यावदिच्छं तथाहते: (?) ।
एकाद्येकोत्तराङ्कानां तावतां हृतिसंयुता ॥
तत एकोनिताद् वृत्या भवेत् सङ्कलनायुति: ।
तद्युक्तिः सुगमा न स्यादिति नेह प्रपञ्च्यते ॥

(The product of any number of natural numbers, beginning with the period and increasing by one, when divided by the product of as many natural numbers beginning with one, will be the repeated sum of the natural numbers, the number of repetitions being one less than the number (of factors in the numerator and

[1]Sri Thomas Heath, *A History of Greek Mathematics*, I, 108-10.

the denominator). The rationale of this will not be easy to understand and so is not detailed here.)

This statement symbolically means Σs_n or the second *saṃkalita*

$$= \frac{n(n+1)(n+2)}{1.2.3}$$

$\sum \frac{n(n+1)(n+2)}{1.2.3}$ or the third *saṃkalita*

$$= \frac{n(n+1)(n+2)(n+3)}{1.2.3.4}$$

$$\sum \frac{n(n+1)(n+2)(n+3)}{1.2.3.4}$$

$$= \frac{n(n+1)(n+2)(n+3)(n+4)}{1.2.3.4.5}, \text{ etc. etc.,}$$

i.e. $\frac{n(n+1)\ldots(n+r-1)}{\lfloor r}$ is the sum of the (r—1)th order of triangular numbers. And the commentator implies that a proof by demonstration similar to the one given above is possible for all these equalities. Only, he does not choose to give it in his commentary written for *alpadhiyām hita*, for the good of the not very intelligent.[1] How one wishes that the learned commentator had not left out the demonstration in this fashion. Already with $\frac{n(n+1)(n+2)}{1.2.3}$ the three-dimensional cube is reached. How will such a demonstration proceed with $\frac{n(n+1)(n+2)(n+3)}{1.2.3.4}$ etc., unless one were to conceive spaces with more than three dimensions ?

9.20.9. The *Yuktibhāṣā* gives a geometrical demonstration for the square of any number n as the sum of the series 1,3,5, . . . to n terms. The method is slightly different from the gnomon method of the Greeks; in fact, it is the method of the *śreḍhīkṣetra*.

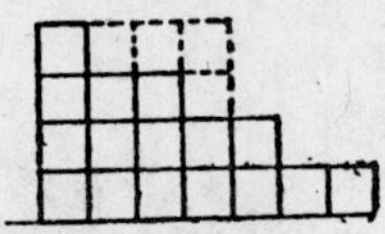

Fig. 37

From the *śreḍhīkṣetra* representing the series of the n natural odd numbers, the portion left in the bottom rows of the figure after the n th small square, is cut off and placed on top of the remainder to complete the square of side n. The method

[1]Verse 4 of the verses of salutation at the beginning of the *Kriyākramakarī*.

is one of induction, the result of observation with *śreḍhīkṣetras* of 2, 3, 4 or 5 terms being generalised.

9.20.10. The difference between the *śreḍhīkṣetra* as conceived by Śrīdhara and Nārāyaṇa and as concieved by the mathematicians of the Āryabhaṭa School is quite obvious. With the mathematicians of the Āryabhaṭa School the *śreḍhīkṣetra* is only a tool for the demonstration of equalities already known (except for the possibility of the demonstration having to deal with multi-dimensional space). But Nārāyaṇa's treatment is more in the nature of an investigation into the possibilities of geometrical treatment of arithmetical progressions. The result is the ability to conceive A.P.s with fractional or negative periods and to attach some meaning to such A.P.s.

This diagrammatic treatment of series seems to be a unique feature of Indian mathematics. True, the Greeks had recourse to the device. They had triangular numbers, square numbers and polygonal numbers, which only means that these numbers can be represented diagrammatically as triangles, squares and other polygons and these figurate numbers are sums of particular arithmetically progressive numbers. But their chief interest was in the numbers themselves, not in the series. According to E. T. Bell[1] the Chinese mathematician Yang-Hiu (126 A. D.) in his *The Analysis of Arithmetical rules in Nine Sections* speaks about the graphic representation of the summation of an A. P. But details of the method are not given.

[1]E.T. Bell—*Development cf Mathematics*, p. 271.

CHAPTER X

SHADOW PROBLEMS AND OTHER PROBLEMS

10.1. Shadow measurements and calculations based on them formed an important part of astronomy and therefore of mathematics from very early times. Even the *Sūryaprajñapti* refers to shadow lengths and their variations according to the time of the day and the year (IV. 9). The highly condensed mathematical section of the *Āryabhaṭīya* devotes more than three verses to problems connected with the gnomon and its shadow :

शंको: प्रमाणवर्गं छायावर्गेण संयुतं कृत्वा ।
यत् तस्य योगमूलं विष्कम्भार्धं स्ववृत्तस्य ॥

(*A.B. Gaṇitapāda*, 14)

(The square of the measure of the gnomon is added to the square of the shadow. The square root of the sum is the radius of the *svavṛtta*);

शङ्कुगुणं शङ्कुभुजाविवरं शङ्कुभुजयोर्विशेषहृतम् ।
यल्लब्धं सा छाया शङ्को : स्वमूलाद्धि ॥

(*A. B. Gaṇitapāda*, 15)

(The distance between the gnomon and the *bhuja* is multiplied by the length of the gnomon and divided by the difference between the gnomon and the *bhuja*. The result is the length of the shadow of the gnomon from its foot.);

छायागणितं छायाग्रविवरमूनेन भाजिता कोटी ।
शंकुगुणा कोटी सा छायाभक्ता भुजा भवति ॥

(*A. B. Gaṇitapāda*, 16)

(The distance between the tips of the shadows multiplied by the length of the shadow and divided by the difference between the two shadows is the *koṭi*. This *koṭi* multiplied by the length of the gnomon and divided by the length of the shadow gives the length of the *bhuja*.)

In interpreting these verses[1] one has to remember that *bhuja* and *koṭi* here are not the base and perpendicular side of the right triangle, but the sine-chord and the cosine-chord and hence the necessity for fixing the circle of reference which, as Nīlakaṇṭha explains, is done in the first of these verses.

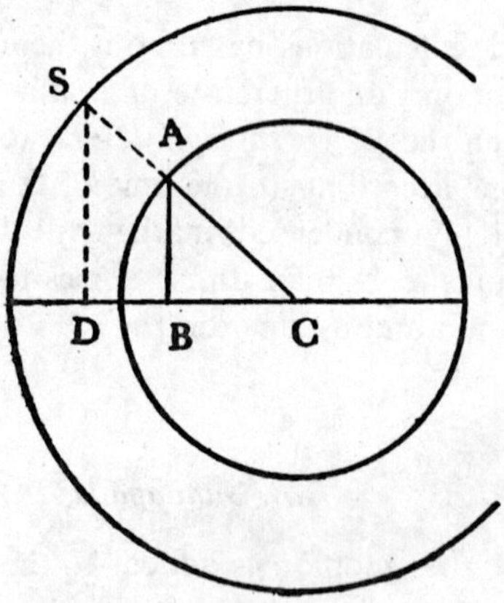

Fig. 1

The source of light S is located on a circle concentric with the *svavṛtta*, (own circle, i.e. the circle of the gnomon and its shadow), which has the tip of the shadow, C as centre and the line joining C to the top A of the gnomon AB as radius. Then SD, the vertical through the source of light is the sine-chord (*bhuja*) and DC is the cosine-chord (*koṭi*).

And the shadow $BC = \dfrac{DB \, . \, AB}{SD - AB}$

(For, from the similar triangles S D C and A B C.

$$\frac{BC}{DC} = \frac{AB}{SD}$$

$$\frac{BC}{DC - BC} = \frac{AB}{SD - AB}$$

$$\text{or } BC = \frac{AB\,(DC - BC)}{SD - AB}$$

$$= \frac{AB \, . \, DB}{SD - AB} \;)$$

The third verse enables us to calculate the height of the source of light and its horizontal distance from the observer with the

[1]W.E. Clark and Rodet (as one gathers from Clark's references to his interpretation) have not been able to interpret these verses satisfactorily.

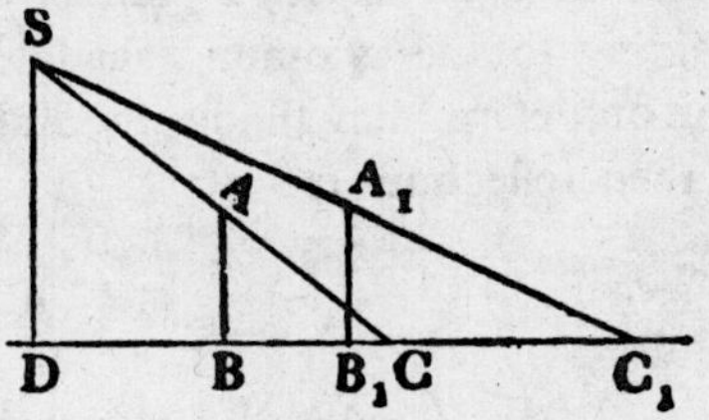

Fig. 2

help of two shadow-throwing gnomons. S is the source of light, AB, and A_1B_1 are two equal gnomons and BC and B_1C_1 their shadows.

Then the *koṭi* i.e. DC or DC_1

$$= \frac{\text{distance between tips of shadows} \times \text{shadow length}}{\text{difference between shadow lengths}}$$

$$= \frac{C\,C_1 \,.\, B\,C \text{ (or } B_1\,C_1)}{B_1\,C_1 - B\,C}$$

and the '*bhuja*' or S D $= \dfrac{\textit{koṭi} \times \text{length of gnomon}}{\text{length of shadow}}$.

$$\text{For } \frac{D\,C}{B\,C} = \frac{S\,D}{A\,B} = \frac{S\,D}{A_1\,B_1} = \frac{D\,C_1}{B_1\,C_1}$$

$$= \frac{D\,C_1 - D\,C}{B_1\,C_1 - B\,C} = \frac{C\,C_1}{B_1\,C_1 - B\,C}$$

$$\text{or } D\,C = \frac{C\,C_1 \,.\, B\,C}{B_1C_1 - B\,C}$$

$$\text{Similarly } D\,C_1 = \frac{C\,C_1 \,.\, B_1\,C_1}{B_1\,C_1 - B\,C}$$

$$\text{Also } S\,D = \frac{A\,B \,.\, D\,C}{B\,C}$$

$$\text{or } \frac{A_1\,B_1 \,.\, D\,C_1}{B_1\,C_1}$$

10.2. Brahmagupta has rules for calculating the time of the day from shadow-measurements, the length of the shadow from the known heights of the gnomon and the light and the horizontal distance between the two (*Br. Sp. Si.* XII., 53) and for finding the height and distance of the light by measuring the shadow lengths of the gnomon at two distances from the light (XII. 54). The last is the same as that given by Āryabhaṭa. The same method is repeated in XIV. 15. How this method of the gnomon and its shadow can be used to measure the height of buildings is explained in the next verse.

10.3. In the chapter *Śaṅkucchāyādijñānādhyāya* (chapter dealing with the knowledge pertaining to the gnomon, shadow etc.) there is an interesting section on determining the height and distance of objects by observing their reflections in water.

युतदृष्टिगृहौच्च्यहृता ह्यन्तरभूमिदृंगौच्च्यसंगुणिता ।
फलभून्यंस्ते तोये प्रतिरूपाग्रं गृहस्य नरात् ॥

(*Br. Sp. Si.*, XIX. 17)

(The distance between the house and the man is divided by the sum of the heights of the house and the man's eyes and multiplied by the height of the eyes. The tip of the image of the house will be seen when the reflecting water is at a distance equal to the above product).

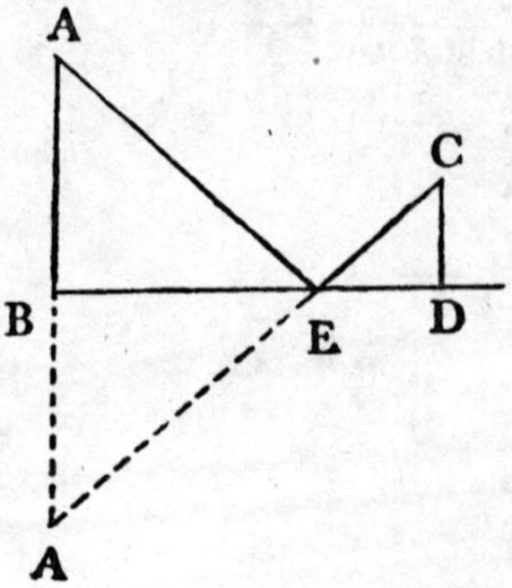

Fig. 3

i.e. if A B is the house (the object), C D the height of the man's eyes and E the reflecting point, the man will be able to see the tip of the image when

$$BE = \frac{BD \cdot CD}{AB + CD}$$

(For, from the similar triangles A B E and CDE

$$\frac{CD}{AB+CD} = \frac{DE}{BE+DE} = \frac{DE}{BD}$$

$$\text{or } DE = \frac{BD \cdot CD}{AB + CD}.\Big)$$

Also from the same pair of similar triangles the height of the house $AB = \frac{BE \cdot CD}{DE}$ (XIX. 18).

By observing the reflection from two different distances also, the height and distance of the object can be determined.

प्रथमद्वितीयनृजलान्तरेणोद्धृता जलापसृतिः ।
दृष्ट्यौच्च्यगुणोच्छ्रायस्तोयान्नृजलान्तरगुणा भूः ॥

(*Br. Sp. Si.* XIX. 19)

(The distance between the first and second positions of the water divided by the difference between the distances of the man from the water, when multiplied by the height of the eyes, is the height, and the same, when multiplied by the distance between the water and the man, is the distance between the water and the house.)

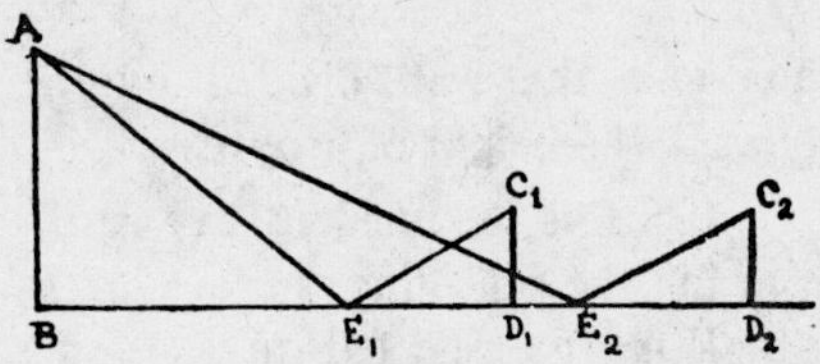

i.e. If A B is the house, $C_1 D_1$ and $C_2 D_2$ the two positions of the observer and E_1, E_2 the two points of reflection.

Fig. 4

$$A\,B = \frac{E_1\,E_2\,.\,C_1\,D_1}{E_2\,D_2 - E_1\,D_1}$$

$$\text{and } B\,E_1 = \frac{E_1\,E_2\,.\,D_1\,E_1}{E_2\,D_2 - E_1\,D_1}$$

$$\text{(For, } \frac{A\,B}{C_1\,D_1} = \frac{B\,E_1}{D_1\,E_1}$$

$$= \frac{A\,B}{C_2\,D_2} = \frac{B\,E_2}{D_2\,E_2}$$

$$= \frac{B\,E_2 - B\,E_1}{D_2E_2 - D_1E_1} = \frac{E_1\,E_2}{D_2E_2 - D_1E_1}$$

$$\therefore A\,B = \frac{C_1\,D_1\,.\,E_1\,E_2}{D_2\,E_2 - D_1\,E_1}$$

$$\therefore B\,E_1 = \frac{D_1\,E_1\,.\,E_1\,E_2}{D_2\,E_2 - D_1\,E_1}\text{)}$$

10.4. A problem combining shadow and reflection is to find the height at which the light from a source at given height reflected from a water-surface between the source and a wall will strike the wall. The problem is posed in

ज्ञातैश्छायापुरुषैर्विज्ञाते तोयकुड्ययोर्विवरे ।
कुड्ये अर्कतेजसो यो वेत्यारूढि स तन्त्रज्ञः ॥

(XIX. 8)

(He is well-versed in devices, who knows how to calculate the ascent of the sun's rays on a wall from the known ratio of the shadow to the object and the distance between the water and the wall) and the solution is contained in

छायापुरूषछिन्नं जलकुड्यान्तरमवाप्तमारूढिः ।

(XIX 20)

(The distance between the water and the wall divided by the ratio of the shadow to the object is the height of ascent.)

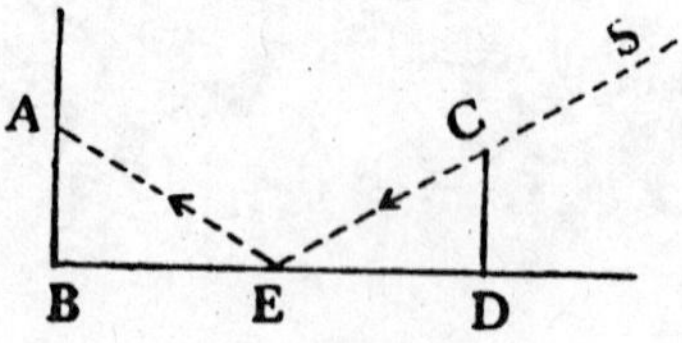

Fig. 5

If S E is the incident ray striking the reflecting surface at E, E A the reflected ray striking the wall at A and C D a gnomon in the path of the incident ray, its shadow will be D E and

$$\frac{AB}{BE} = \frac{CD}{ED}$$

$$AB = BE \cdot \frac{CD}{ED} = \frac{BE}{\frac{DE}{CD}}$$

More than the possibility of Brahmagupta having investigated the laws of reflection, the solutions of these problems testify to a familiarity with the operations like alternando and invertendo, connected with proportion and hence with a more or less thorough grasp of the theory of proportionality.

10.5. Śrīdhara has rules for calculating the time of the day from the length of the shadow and vice versa. Mahāvīra gives the time-honoured method of fixing the cardinal directions, explained in the *Śulba sūtras*. The time of the day is calculated with the help of the formula :[1]

$$\text{Time elapsed} = \frac{1}{2\,(s/g+1)} = \frac{1}{2\,(\cot A + 1)}$$

where s is the shadow length, g the length of the gnomon and A the altitude of the sun. This is strictly applicable only when $A = 45^0$. When the equinoctial shadow is not zero the formula has to be modified into[2]

[1]*G.S.S.* IX. $8\frac{1}{2}$.
[2]*G.S.S.* IX. $15\frac{1}{2}$.

Time elapsed $= \dfrac{g}{2(s+g-e)}$ where e is equinoctial shadow, and the shadow length at any time of the day is given by $s = \dfrac{g}{2d} - g + e$ where d is the fraction of the day elapsed or remaining.[1] With the ratio of the shadow to the gnomon length known, the length of the shadow of a vertical gnomon intercepted by a vertical wall in front can be calculated with the help of similar triangles.[2] With the help of this formula, when the shadow height on a vertical screen is known, the height of the object casting the shadow or the distance between the object and the screen can be calculated.[3]

10.6. An unusual problem is the calculation of the inclination (*avanati*) of an inclined object with the help of the length of the shadow cast by it.

छायावर्गाच्छोध्या नरभाकृति गुणितशङ्कुकृतिः ।
सैकनरच्छायाकृतिगुणिता छायाकृतेः शोध्या ॥
तन्मूलं छायायां शोध्य नरभानवर्गरूपेण ।
भागं हृत्वा लब्धं स्तम्भस्यावनतिरेव स्यात् ॥

(*G. S. S.* IX. 32-33)

(From the square of the shadow length should be subtracted the square of the pillar length multiplied by the square of the ratio of the shadow to the object. The difference multiplied by the square of the ratio increased by one is again subtracted from the square of the shadow (of the pillar). The square root of this difference subtracted from the shadow length and divided by the square of the shadow-object ratio as combined with one, gives the inclination of the pillar.)

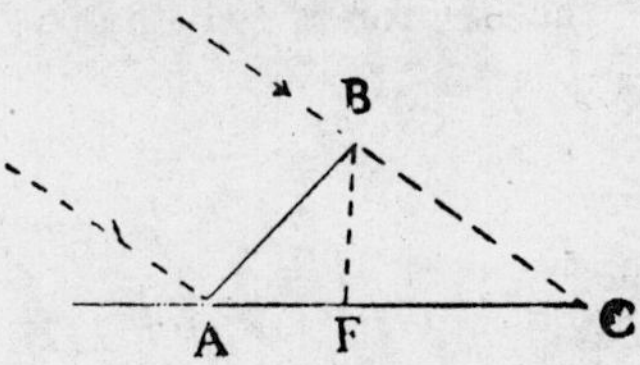

Fig. 6

Here by *avanati* (inclination) is meant the projection of the slanting object on the horizontal, AB is the slanting pillar casting the shadow AC on the horizontal plane. B F is perpendicular on AC.

[1]Ibid IX. 18.
[2]Ibid IX. 21.
[3]Ibid IX. 23 and 26.

$$\text{Then } AF = \frac{AC - \sqrt{AC^2 - (AC^2 - s^2 . AB^2)(s^2+1)}}{s^2+1}$$

where s is the shadow-object ratio.

For CF, which is the shadow of the vertical BF

$$= BF . s = s\sqrt{AB^2 - AF^2}$$

$$\text{Also} \quad CF^2 = (AC - AF)^2$$

$$= (BF . s)^2 = s^2 (AB^2 - AF^2)$$

$$\text{i.e. } AC^2 + AF^2 - 2 AC . AF = s^2 . AB^2 - s^2 . AF^2$$

$$\text{or } AF^2 (s^2 + 1) - 2 AC . AF = s^2 \times AB^2 - AC^2$$

$$\text{or } AF^2 - \frac{2 AC . AF}{s^2+1} = \frac{s^2 . AB^2 - AC^2}{s^2+1}$$

$$\text{i.e.} \quad \left(AF - \frac{AC}{s^2+1}\right)^2 = \frac{s^2 . AB^2 - AC^2}{s^2+1} + \frac{AC^2}{(s^2+1)^2}$$

$$\therefore AF = \frac{AC}{s^2+1} - \frac{\sqrt{AC^2 + (s^2 . AB^2 - AC^2)(s^2+1)}}{s^2+1}$$

$$= \frac{AC - \sqrt{AC^2 + (s^2 . AB^2 - AC^2)(s^2+1)}}{s^2+1}$$

The section concludes with the simple calculation of the length of the shadow, the height of the source of light, and its distance from the gnomon when the other two are known.[1]

Āryabhaṭa II's treatment of shadow problems is confined to the calculation of the time of the day from the shadow and vice versa.

10.7. The shadow problems in the *Līlāvatī* are purely geometrical and evidently modelled on the elder Āryabhaṭa's treatment. The solution of the first problem in the section is worthy of notice.

The rule is

छाययो : कर्णयोरन्तरे ये तयोर्वर्गविश्लेषभक्ताः रसाद्रीषवः : ।
सैकलब्धे : पदघ्नं तु कर्णान्तरं भान्तरेणोनयुक्तं दले स्तः प्रभे ॥

(*Līl.* 232)

(576 is to be divided by the difference between the squares of the difference of the shadow lengths and that of the hypotenuses. The difference between the hypotenuses multiplied by the square

[1] Ibid. IX $40\frac{1}{2}$-45.

root of the above quotient, combined with one, is to be combined with and diminished by the difference between the shadow-lengths. The results when halved give the shadow-lengths separately.)

The figure attached to this problem does not represent the shadows cast by two equal gnomons at two different distances from the source of light but a triangle formed by the juxtaposition of two *jātyas* (right triangles) (ABC and ABD in the figure), i.e., as if two sources of light at different heights or distances or both on either side of the gnomon were producing shadows BD and BC on either side of the gnomon. Then these are the *ābādhās* of the triangle ACD. Let the difference between the shadow-lengths, BD − BC = a and difference between the hypotenuses, AD−AC = b be given.

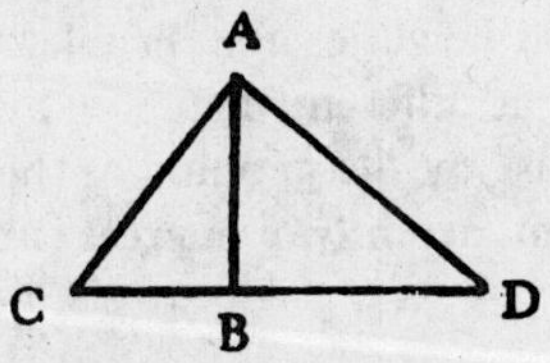

Fig. 7

Let BD+BC = x

$$\text{Then}\quad BD = \frac{x+a}{2} \text{ and } BC = \frac{x-a}{2}$$

$$AD^2 - AC^2 = BD^2 - BC^2 = ax$$

$$\therefore\ AD + AC = \frac{a\,x}{b}$$

$$\therefore\ AD = \frac{ax/b + b}{2} = \frac{a\,x + b^2}{2\,b}$$

$$\text{and}\quad AC = \frac{a\,x - b^2}{2\,b}$$

$$\text{But } AD^2 = AB^2 + BD^2$$

$$\therefore \left(\frac{ax + b^2}{2\,b}\right)^2 = AB^2 + \left(\frac{x+a}{2}\right)^2$$

$$a^2 x^2 + 2ab^2 x + b^4 = b^2 (4\,AB^2 + x^2 + 2ax + a^2)$$

$$x^2 (a^2 C b^2) = b^2 (4\,AB^2 + a^2 - b^2)$$

$$x^2 = b^2 \cdot \frac{4\,AB^2}{a^2 - b^2} + 1$$

$$x \text{ or } BD + BC = b\sqrt{\frac{4\,AB^2}{a^2 - b^2} + 1}$$

$$\text{Hence BD or BC} = b\,\frac{\sqrt{\dfrac{4\,AB^2}{a^2-b^2}+1}\pm a}{2}$$

And this is the expression embodied in the verse. Since AB = 12 angulas, 4 AB^2 appears as 576.

All the other problems are solved with the help of similar triangles.

Nārāyaṇa has nothing new except that he varies the problem of the shadows cast by the gnomon at different distances by calculating the lengths of the shadow cast by the gnomon at the same place when it intercepts the light from two sources at different heights on the same vertical.[1]

[1]*G.K.* p. 211.

GLOSSARY OF GEOMETRICAL TERMS

aṃsa	— an upper vertex of a quadrilateral. (*Sl. Su.*)
akṣṇayārajju	— the diagonal chord of rectangle or a square (*Sl. Su.*).
ābādhā, avabādhā, avadhā, abadhā or badhā	— segments of the base of a triangle produced by the altitude on it; the projection of any slanting side on the horizontal.
āyata or āyatacaturbhuja	— rectangle. Syn. — āyatacaturasra, dīrghacaturasra or dīrgha (*Sl. Su.*).
āyatavṛtta	— Ellipse (*G.S.S.* VII. 5).
āyāma	— Breadth (*A. B. Gaṇitapāda*) Nīlakaṇṭha's commentary on A.B. 8; length as opposed to breadth, (*Jambudvīpasamāsa* pp. 5,7 and other Jain works and later works).
utsedha	— height. Syn.—ucchraya, ucchriti, auccya.
udīcī	— north-south line.
ubhayataḥprauga	— double isosceles triangle; rhombus (*Sl. Su.*).
ṛjubhuja	— rectilinear figure (*Līl.* 163).
karaṇī	— the side of a rectilinear figure; the side of a square or rectangle (*Sl. Su.*); maker or producer of the required area.
karṇa	— diagonal; hypotenuse. meaning in the *Katyāyana Śulba Śūtra* (lv, 9 and 10) is obscure. Syn.—śruti, śravaṇa śravas (*G.K.*; *Ks. Vya.* 26, 27 etc.), śrọtra (*G.K.*; *Ks. Vya.* Ud. 93)
koṭi	— the perpendicular side of a right-angled triangle.

koṭijyā — the cosine-chord of an arc, i.e. the half-chord of its complementary arc.

koṇa — angle, corner.

kṣetra — closed figure.

kṣetragaṇita — geometry.
Syn.—Bhūgaṇita (*G.K.*, *Ks. Vya. Ud.* 90). Bhūmigaṇita (*G.K.K. Vya.* 132).

kṣetraphala — area.

khātaphala — the volume of a pit or excavation

gaṇita — area, mathematics, astronomical calculations.

grāsa — the common portion of two intersecting circles; the largest witdh thereof.

goḷa — sphere.

ghanaphala — volume.

ghātakṣetra — the diagrammatic representation of a multiplication product.

cakravālavṛtta — annulus (*G.S.S.* VII. 6.).

caturasra — quadrilateral square (*Sl. Su.*).

catuṣkoṇa — quadrilateral (*S.P.*I. 8 *K.K.*).

cāpa — arc.
Syn.—Dhanus, Kārmuka, Kodaṇḍa.

cāpakṣetra — segment.

chāyā — shadow. Syn.—Bhā, Prabhā.

janya — rational right triangle or rectangle from which other rational figures are to be obtained, a figure with rational sides (*G.S.S.* VII. 90½ and onward).

jātya — a rational right angled triangle, any rectilinear figure with rational sides.

jyā — chord, Syn.—Jīvā, Maurvī, Śiñjinī (*G.K.*, *Ks. Vya.* 12), Guṇa (*G.K.*, *Ks. Vya. Ud.* 57).

tiryaṅmānī — the transverse side of a quadrilateral; the horizontal side. The transverse measurer (*Sl. Su.*).

trikarṇa — (*K. Sl.* IV, 9) meaning not clear.

trijyā — radius; the sine-chord of 3 *rāśis* or of one-fourth the circumference.

tribhuja — triangle Syn.—Tribāhu, trikoṇa (*G.K. Ks., Vya.* 13, 27 etc.), tryasra.

trisama — equilateral triangle, trapezium with three sides equal.

tryasra — triangle, more especially a right triangle.

dvisamabhuja or dvisama — isosceles.

dhanuḥkāṣṭha — arc; bow-stick (literally) (*G.S.S.* VII. 73½).

nara — gnomon. Syn.—nṛ.

nirgama — annulus (*G.K., Ks. Vya.* 11; *G.S.S.* VII 28).

nemi — a part of an annulus (*G.K. Ks. Vya*, 14; *G.S.S.* VII. 7).

parakarṇa — the third diameter of a cyclic quadrilateral obtainable by interchanging a pair of adjacent sides. (*G.K., Ks. Vya.* 48, 96).

paridhi — circumference. Syn. —pariṇāha, vṛti.

parimaṇḍala — ellipse (?) (*S.P.*) I.8; Circle.

pāta — the circumcentre-cum-incentre-cum-orthocentre of an equilateral triangle (Nīlakaṇṭha's comments on *A.B., Gaṇitapāda* 6).

pātarekhā — the perpendiculars on the base and top of a trapezium from the point of intersection of its diagonals. (*A.B., Gaṇitapāda.* 8).

pārśvamānī — the lateral side of a quadrilateral;

	the flank-side; the flank-measurer (*Sl. Su.*).
pṛṣṭhaphala	— surface area.
pṛṣṭhyā	— line of symmetry (*Sl. Su.*) usually east-west line.
prauga	— triangle, an isosceles triangle (*Sl. Su.*); the forepart of the shaft of a chariot, which is triangular in shape.
bhuja	— sides. Syn.—bāhu, dos
bhujajyā	— sine-chord of an arc, i.e. half-chord of twice the arc.
bhū	— base Syn.—mahī, ku, vasudhā, urvī, tala bhumī, dharā.
maṇḍala	— circle (*Sl. Su.*).
madhyamalamba	— mean altitude (*G.K.Ks. Vya.* 63 and *Ud.* 54).
mukha	— face, the top side of a figure with more than three sides, especially the top or the shorter parallel side of a trapezium. Syn.—Vadana; Vaktra (*G.K., Ks. Vya. Ud.* 18)
rajju	— perimeter (*G.K., Ks. Vya. Ud.* 84,85 and 86; *G.S.S.* VII. 38).
ruṇḍa	— breadth (of an annular ring) (*Trilokasāra* III. 315)
lamba or avalamba	— perpendicular; altitude; vertical.
valayākārakṣetra	— figure shaped like a ring, annulus.
varga	— a small square of unit side got by dividing the sides into units and drawing parallels through the points of division (*Ap. Sl. Su.* III. 7 and other *Sl. Su.*).
viṃdaphala	— volume (*T. P.* I. 181).
viśeṣa	— the difference between the diagonal and side of a square, especially when expressed in terms of the side (*Sl. Su.*).
viṣkambha	— diameter, especially in *Sl. Su.* A.B.

	and Jaina works; breadth as opposed to length (*Jambudvīpasamāsa* p. 5. 7 and other early Jaina works).
viṣamacakravāla	— ellipse (?) (*S. P.* I. 8).
viṣama or Viṣama-caturbhuja	— a quadrilateral with unequal sides; a cyclic quadrilateral (in the *Br. Sp. Si.* etc.) Syn. — viṣamacaturasra, viṣamabhuja.
viṣamatribhuja	— scalene Triangle.
vistāra	— length (Nīlakaṇṭha's commentary on *Āryabhaṭīya* under v. 8); breadth.
vistṛti	— diameter (*Y. B. p.* 208).
vṛti	— perimeter (*G. K. Ks. Vya.* 112).
vṛtta	— circle Syn. — Valaya, Maṇḍala (*Sl. su.*) (*T. P.*, *Trilokasāra*).
vedha	— depth.
vyāsa	— diameter; breadth (*Jambudvīpa-samāsa* p. 20; *Trilokasāra*, 310).
vyāsārdha	— radius Syn. — Viṣkambhārdha.
śaṅkhavṛtta	— a figure roughly resembling the longitudinal section of a conch shell (*G.K. Ks. Vya.* 2, 11, 12.).
śaṅku	— gnomon.
śara	— arrow, the height of an arc or segment of a circle. Syn. — Iṣu; bāṇa: sāyaka (*G. K. Ks.* 66).
śṛṅgāṭaka	— triangle (Monier Williams); some sort of a four sided figure (*Ma. Si.* XIV. 74, 79) tetrahedron [*G. S. S.* VIII. 30½ (?)].
śreḍhīkṣetra	— diagrammatical representation of a mathematical series.
śroṇi	— a lower vertex of a quadrilateral or triangle (*Sl. Su.*).
ṣaḍaśri	— tetrahedron (*A. B. Gaṇitapāda.* 6)
samakoṣṭhamiti	— area, the measure of equal unit squares in a figure (*Līl.* 167).
samacaturbhuja	— square or rhombus; a quadrilateral

with all four sides equal. Syn.—Tulyacaturbhuja.

samabāhu — equilateral figure (*G. K. Ks. Vya.* 1).

samānalamba — a quadrilateral with the altitudes equal; trapezium. Syn.—Samalaṃba.

samacakravāla — circle (*S. P.* I. 8).

samadalakoṭi — altitude (?) (*A.B. Gaṇitapāda* 6).

samapariṇāha — the circumference of a circle (*A.B. Gaṇitapāda*. 7).

saṃpāta — point of intersection (*A. B. Gaṇitapāda* 18 and Nīlakaṇṭha's commentary on it).

sandhi — the interspace between the foot of an altitude and the foot of the flank side from whose tip the altitude is drawn, usually in a quadrilateral. (*Līl.* 193).

sūcī — the inner, outer or middle, diameter of an annular ring (*Trilokasāra*, III, 309, 310).

sūcī or sūcīkṣetra — the triangle got by producing the flanks of a quadrilateral till they meet; the pyramid or cone got by producing the lateral faces of the frustum of a pyramid or cone.

hṛdaya or hṛt or hṛdayarajju — circum-radius (*Br. Sp. Si.* XII, 27; *G. K. Ks. Vya. Ud.* 48).

BIBLIOGRAPHY

1. *Sanskrit and Prakrit Works*

1. Anuyogadvārasūtram with Hemacandra's Maladhārīya. Jainagranthamālā, Bhavanagar. 1939.
2. Āpastamba Śulbasūtra with the commentaries of Kapardisvāmin, Karavinda and Sundararāja. Ed. by D. Srinivasachar and S. Narsimhachar. University of Mysore-Oriental publications. Sanskrit series No. 73. 1931.
3. Āryabhaṭīya with the Bhāṣya of Nīlakaṇṭha Somasutvan, Gaṇitapāda. Ed. by K. Sambasiva Sastri. Tvm. Skt. Series No. C 8. 1930.
4. Āryabhaṭīya with the commentary Bhaṭadīpikā of Paramādīśvara. Ed. by Kern. Leiden. 1875.
5. Atharvaveda Saṃhita (Śaunakīya) with Sāyaṇa's commentary and a Hindi Translation. Ed. by Pandit Ramchandra Shastri. Sanatana Dharma Press, Moradabad.
6. Bakhshali Manuscript—A study in medieval Mathematics. Ed. by G. R. Kaye, Calcutta 1927, 1933.
7. Bhagavatīsūtram. Āgamodayasamiti. 1921.
8. Bhāskariyabījagaṇitam Ed. by V. G. Apte. Ānandāśrama Samskṛta Granthāvalī-99. 1930.
9. Brahmasphuṭa Siddhānta and Dhyānagrahopadeśa of Brahmagupta. Ed. with notes by Sudhakara Dvivedi. Reprint from the Pandit. 1902.
10. Gaṇita Kaumudī of Nārāyaṇa Paṇḍita Parts I & II. Ed. Padmakara Dvivedi. The Princess of Wales Sarasvatī Bhavana T.xts. No. 57. Banaras.
11. Gaṇita Sārasaṁgraha of Mahāvīra Ed. with translation and notes by M. Rangacarya. Madras Government publication. 1912.
12. The Gommaṭasāra of Nemicandra, Jīva Kāṇḍa Ed. with introduction, Translation and commentary by J. L. Jain. The Sacred Books of the Jainas. Vol. V Lucknow.
13. Jambudvīpaprajñaptisaṃgraha. Ed. A. N. Upadhye and Hiralal Jain with introduction by Lakshmichandra Jain—Jaina Samskṛti Samrakṣak Sangh, Sholapur. 1958.

14. Jambudvīpasamāsa of Umāsvāti. Satyavyayagranthamālā No. 2. Ahmedabad. 1923.

15. Jyotiṣkaraṇḍaka of Vālabhyācārya with Malayagiri's commentary. Rishabhadevji Kesarimalji Samstha, Rutlam. 1928.

16. Karaṇa Paddhati Ed. by K. Sambasivasastri. Tvm. Sans. Series No. c XXXI.

17. Karaṇa Paddhati. Ed. by Dr. S. K. Nayar with an introduction. Madras University publications. 1956.

18. Kātyāyana Śulba Pariśiṣṭa (Part only). Pandit, New Series Vol. 4. 1882.

19. Līlāvati with Buddhivilāsinī of Gaṇeśa and the Vivaraṇa of Mahīdhara. Ānandāśrama Samskṛta Granthāvali, Poona. 1937.

20. Mahābhāskarīya of Bhāskara I with the Bhāṣya of Govindasvāmin and the super commentary Siddhānta-dīpikā of Parameśvara. Ed. with introduction and notes by T. S. Kuppanna Sastri. Government Oriental Manuscript Library, Madras, 1957.

21. Mahāsiddhānta of Aryabhaṭa II. Ed. with notes by Sudhakara Dvivedi, Banaras 1910.

22. Maṇḍapadruma of Mahādevasūri. Ed. by Sree Krishna Sharma, Adyar Library Bulletin, Vol. XXII. 1958, pp, 119-157.

23. Pāṭigaṇita of Śrīdharācarya. Ed. K. S. Shukla, Lucknow University, Hindu Astronomical and Mathematical Text Series, No. 2. 1959.

24. Prajñāpanopāñgam with Malayagiri's commentary. Nirnayasagar Press, Bombay.

25. Ṛg Veda Saṁhitā with Sāyaṇa's Commentary. Vaidika-saṁśodhana Maṇḍala of the Tilak Maharastra Vidya Pith, Poona. 1933.

26. Śatapatha Brāhmaṇa (Mādhyandinaśākhīya). Ed. by Candradhara Sharma. Achyutagranthamala, Kasi.

27. Śatapathabrāhmaṇa. Eggeling's Translation. S.B.E. Vol. 43.

28. Ṣaṭkhaṇḍāgama with Vīrasena's Dhavalā Tīkā (part IV). Ed. by Hiralal Jain with an article on the Mathematics of

the Dhavaḷā by A. N. Singh. D. Sc. Jaina Sahityoddharaka Fund, Amaraoti.

29. Siddhānta Śekhara of Śrīpati—Sanskrit Astronomical work of the Eleventh Cent. Ed. with the commentary of Makhibhaṭṭa (Ch. IV. X) and an original commentary (Ch. IV) by Babuaji Misra Part I, Calcutta University Press.

30. Siddhāntatattvaviveka of Kamalākara. Ed. with Śeṣavāsanā by Sudhakara Dvivedi. Banaras. 1885.

31. Siddhānta Śiromaṇi of Bhāskara II. Ed. by Bapudeva Sastri and revised by Pandit Chandra Deva, Banaras. 1891.

32. Siddhānta Sārvabhauma of Munīśvara. Ed. by Gopinath Kaviraja. The Princess of Wales Saraswati Bhavana texts, No. 41. 1932. Banaras.

33. Śrauta Diagrams, drawn by a priest who has performed the sacrifices and kept in the Govt. Or. Mss. Lib. Madras.

34. The Śulbasūtra of Baudhāyana with the commentary of Dvārakānātha Yajvan called Śulbadīpikā. Ed. by G. Thibaut. The Pandit (Banaras) Old Series IX and X, 1874-5 and New Series I. 1877.

35. Śulbasūtram of Kātyāyana. Ed. by Vidyadhara Sharma. Achyuta Granthamala, 3, Kasi. 1928.

36. Sūryaprajñaptyupāñgam with Malyagiri's commentary. Agamodaya-Samiti, Bombay. 1929.

37. Taittirīya Brāhmaṇa with Bhaṭṭabhāskara's commentary. Mysore Government Oriental Library Series, Bibliotheca Sanskritica No. 36. 1908.

38. Tantrasaṁgraha of Nīlakaṇṭha with the commentary of Tṛkkuṭavelil Śankara Vāriyar. Ed. by Śūranad Kuñjan Pillai, Tvm. Sans. Series, No. 188. 1958.

39. Tantrasamuccaya of Nārāyaṇa with the commentary, Vimarśinī of Śankara. Ed. by Ganapati Sastri. Tvm. Skt. Series No. LXVII.

40. Tattvārthādhigamasūtra of Umāsvāti. Ed. by Khubcandraji Siddhanta Shastri, Raycandra Jaina Sastramala.

41. Tiloyapaṇṇathi of Yativṛṣabhācārya. Ed. by A. N.

Upadhye and Hiralal Jain. Jaina Samskrti Prakasaka Sangha, Sholapur 1943.

42. Trilokasāra of Nemicandra. Ed. by Manoharlal Shastri. Manikchand Digambarlal. Jainagranthamāla. 12.
43. Triśatikā of Śrīdharācārya. Ed. by Sudhakara Dvivedi. Banaras 1899.
44. Yuktibhāṣā Part I. Ed. with notes by Ramavarma (Maru) Thampuran and A. R. Akhileswara Aiyar. Mangalodayam Ltd., Trichur. 1952. (1128 M. E.)

II Manuscripts

1. Bhāskarakṛtam Āryabhaṭīyabhāṣyam. R. 14850. Govt. Or. Mss. Lib. Madras
2. Kriyākramakarī, a transcript of the Manuscript (no. R. 2754) in the Govt. Or. Mss. Lib. Madras in the possession of Sri K. V. Sharma.
3. Kuṇḍadarpana, R. 349. Govt. Or. Mss. Lib, Madras.
4. Līlāvatī-vyākhyā of Parameśvara. R. 5160. Govt. Or. Mss. Lib., Madras.
5. Mānava Śulbasūtra with Śivadāsa's Bhāṣya I. E, 17. Asiatic Society Library, Calcutta.
6. Sadratnamāla R. 4448. Govt. Mss. Lib., Madras.
7. Siddhānta Śekhara of Śrīpati, D. 13426. Govt. Or. Mss. Lib., Madras.
8. Tantrasaṃgraha-vyākhyā, T. 275 (572). The Sanskrit College Library, Trippunithura.
9. Tantra-Saṃgraha-Vyākhyā. A Transcript in the keeping of Ramavarma (Maru) Thampuran of a manuscript in the Deśamangalattu Mana.

III Modern Works in English and Indian Languages

1. Banerjee H. C.—Colebrook's translation of Līlāvatī, with notes.
2. **Bell** E. T. The Development of Mathematics. Second Edition. 1945 New York London (McGraw Hill Book company).
3. Bower C. B.— History of Calculus, Dover 1959.
4. Būrk Albert Das—Āpastamba Śulba Sūtra, Ed. with trans-

lation and introduction. Z. D. M. G., LV, 1901 pp. 543-91 and LVI, 1902 pp. 327-39.

5. Cajori F. A History of Mathematics, New York, 1919.
6. Cantor Moritz. Vorlesungen Uber Geschichte der Mathe-Matik, Vol. I, 4th Edition.
7. Clark Walter Eugene. The Āryabhaṭīya, translated with notes, The University of Chicago Press, Chicago. 1930.
8. Colebroke H. T. Miscellaneous Essays. Vol. II.
9. Datta B. B. The Science of the Sulba. University of Calcutta, 1932.
10. Datta B. B. and Singh A. N. History of Hindu Mathematics, Parts I, II. Lahore. 1935 and 1938.
11. Dickson L. E. History of the Theory of Numbers, Vol. II Diophantine Analysis.
12. Dixit S.B. Bhāratīya Jyotiṣa-Hindi Tianslation by Sivanath Jharkhandi. Publications Division, Uttar Pradesh Government, 1957.
13. Dvivedi Sudhakara. (i) Gaṇakataraṅgiṇi—Banaras 1933. (ii) A History of Mathematics I, Banaras 1910.
14. Hogben Lancelot. Mathematics for the Million. New Edition. 1957. George Allen & Unwin Ltd, London.
15. Heath, Sir Thomas. A History of Greek Mathematics, Oxford, 1921.
16. Hogben Lancelot. Mathematics in the Making, London 1960.
17. Kaye G.R. (i) Hindu Astronomy Memoirs of the Archaeological Survey of India, No. 18. 1924. (ii) Indian Mathematics. Thacker, Spink and Co. Calcutta and Simla 1915.
18. Needham Joseph. Science and Civilisation in China Vol. 3. Mathematics. Cambridge, 1959.
19. Neugebauer. The Exact Sciences in Antiquity. Princeton. 1952.
20. Parameśvara Iyer, Ullur S. Keraḷa Sāhitya Caritram, Parts II and III.
21. Premi Nathu Ram—Jaina Sahitya aur Itihasa. Bombay, 1942.

22. Renou, Louis—Vedic India. Translated by Philip Sprat, Calcutta.
23. Sachau Edward C. Alberuni's India, edited with notes and indices. Delhi. 1964.
24. Smith D. E. The History of Mathematics, 1925. Ginn and Co. Boston.
25. Smith D. E. and Mikami Yoshio. History of Japanese Mathematics. Chicago. 1914.
26. Thibaut G. Astronomie, Astrologie and Mathematik, Strassburg. 1899, belonging to the series. "Grundriss der indoarischen Philologie und Altertumskunde, iii Band, 9 Heft.
27. Vaṭakkumkur Rāja Rāja Varma Rāja. Keralīya Samskṛta Sāhitya Caritram. Parts I, II and III, Mangalodayam, Trichur 1937, 1946 and 1949.

N. B. References to the Siddhāntaśekhara by page number are to the Manuscript and by verse number to the printed edition.

INDEX